Henri Ulrich

Raw Materials for Industrial Polymers

Henri Ulrich

Raw Materials for Industrial Polymers

Hanser Publishers, Munich Vienna New York

Distributed in the United States of America by
Oxford University Press, New York
and in Canada by
Oxford University Press, Canada

Dr. Henri Ulrich
Director,
Dow Chemical USA
North Haven Laboratories
North Haven, CT.

Distributed in USA by
Oxford University Press
200 Madison Avenue, New York, N.Y. 10016

Distributed in Canada by
Oxford University Press, Canada
70 Wynford Drive, Don Mills, Ontario M3C IJ9

Distributed in all other countries by
Carl Hanser Verlag
Kolbergerstraße 22
D-8000 München 80

Library of Congress Cataloging-in-Publication Data

Ulrich, Henri. 1925-
Raw materials for industrial polymers / Henri Ulrich.
p. cm.
Bibliography: p.
Includes index.
ISBN 0-19-520762-9 (U.S.)
1. Polymers and polymerization. I. Title
TP1087.U45 1988
668.9--dc19 88-16256
CIP

CIP-Titelaufnahme der Deutschen Bibliothek

Ulrich, Henri:
Raw materials for industrial polymers/Henri Ulrich. – Munich ; Vienna ; New York : Hanser ; New York : Oxford Univ. Press, 1988
ISBN 3-446-15099-4 (Hanser)
ISBN 0-19-520762-9 (Oxford Univ. Press)

ISBN 3-446-15099-4 Carl Hanser Verlag, Munich, Vienna, New York
ISBN 0-19-520762-9 Oxford University Press, New York
Library of Congress Catalog Card Number 88-16256

Printed in the Federal Republic of Germany by Spandel-Druck, Nürnberg

Preface

Presently about 85% of the basic industrial chemicals are used in the production of industrial polymers, and most of the organic intermediates are based on oil and natural gas. Annual worldwide polymer consumption reached approximately 66 million metric tons in 1979 but has been declining slowly since. The decline was prompted by an increase in feedstock prices brought on by the 1979 oil crisis. In view of the recent oil glut petrochemical feedstocks will dominate the industry for the foreseeable future; however, it is expected that coal will play a more important role in the next century.

This book summarizes the present state of technology in monomer production as well as future trends. It is intended as a companion volume to my earlier book *Introduction to Industrial Polymers* (Hanser, Munich, New York, 1982). The methods of synthesis used in the manufacture of major polymer raw materials are described, and new emerging technologies are highlighted. The relationship of major polymers to basic feedstocks such as syngas, methanol, ethylene, propylene, benzene, toluene, and xylenes is indicated. Data on world consumption and current prices of the basic building blocks are given. A very important factor in the usefulness of an industrial polymer is its price, and the economics and availability of the monomer become critical issues. Today we are witnessing a trend toward C_1 chemistry that in the long run will make basic building blocks such as methanol, formaldehyde, and acetic acid available at lower cost.

In order to emphasize feedstock considerations this book is organized around the number of carbon atoms in the feedstock. Chapters 2 to 6 deal with the aliphatic building blocks with up to five carbons. Chapters 7 to 9 describe the aromatic building blocks as well as aliphatic monomers obtainable from the BTX (benzene, toluene, xylene) fraction available from oil refineries. Monomers for high performance polymers are discussed in Chapter 10, and information on monomers and polymers obtainable from renewable resources is summarized in Chapter 11. The final chapter deals with future trends in the development of special monomers to further advance the state of industrial polymers as viable materials of construction as well as functional materials in tomorrow's world.

We have already witnessed the emergence of the plastic artificial heart. New automobiles made entirely with nonrusting plastic body panels are already sold in the United States. Reinforced plastics are used increasingly in the production of modern aircraft. Fiber optics and conductive polymers will make an impact on the electronics business, and membrane technologies will aid in the desalination of seawater and in the easy separation of industrial gases. High performance polymers will be the materials of choice in the construction of solar sails that will propel future space vehicles. Functional polymers similar in sophistication to the polyproteins, the elements of life, are emerging, with genetic engineering employed in their construction. It is indeed exciting to participate in the plastics revolution, which will clearly continue into the next century.

Guilford, Connecticut
January 1988

Henri Ulrich

Contents

Chapter I. General Introduction

1. Historical Overview

The chemical industry as we know it began in 1824 when Wöhler discovered that the organic natural product urea could be synthesized from inorganic matter. The new science of synthetic organic chemistry was enthusiastically practiced at university laboratories in Europe, the foremost of which was Liebig's laboratory at the University of Giessen in Germany. Soon afterwards industrial organic chemicals such as aniline and phenol were isolated from coal tar, which was readily available from the gas-producing plants in Europe. (Coal-derived gas was needed to light the cities in the preelectric age). The development of the chemical industry was closely associated with the production of synthetic dyestuff based on coal tar. The first synthetic dyestuff factories opened in Europe around 1830. New discoveries led to new industrial applications. In 1863 Nobel applied for his first explosives patent, and in 1873 Kolbe's synthesis of salicylic acid led to the establishment of the synthetic drug industry.

Industrial polymers emerged at the beginning of this century. Leo Hendrik Baekeland discovered phenol formaldehyde resins in 1909. These materials, which became known as Bakelite, were used widely in the manufacture of molded parts, including the first mass-produced telephones. Today telephones are made from acrylonitrile/butadiene/styrene terpolymers (ABS). Cellulose acetate and poly(vinyl chloride) (PVC) were invented in 1927, and the important polyolefins such as polystyrene (1930), polyethylene (1938), and polypropylene (1953) followed suit. W. H. Carothers synthesized the first condensation polymer (nylon-6,6) in 1935, and the polyesters (PET) appeared several years later in 1941. Engineering thermoplastics such as polycarbonate (1953), poly(phenylene oxide) (1956), and polyacetal (1958) are relative newcomers on the scene.

By 1959 approximately 2.7 million metric tons of plastics were produced in the United States, and by 1984 this number had reached 17.5 million metric tons. Man-made fibers added another 4.3 million metric tons, and 2 million metric tons of synthetic rubber was also produced in the United States in 1984. All of these industrial polymers are made from the basic building blocks described in this book.

Since virtually all of the organic chemicals required by industry can be obtained from carbon and water, biomass, animal products, coal, oil, and natural gas have all been considered as raw materials. The choice of feedstock depends always on availability, cost, and conversion technology. In the early days of the chemical industry, organic chemicals were produced by fermentation. However, this method was practical for only a limited number of products such as ethanol and dilute acetic acid (vinegar). Coal was the raw material base until World War I. Since crude oil has a hydrogen-to-carbon ratio of 1.8:1, which is superior to that of coal's 0.8:1, it soon became the feedstock of choice for synthetic organic chemicals.

Crude oil also has a higher caloric value than coal and does not produce ash on burning. Therefore, it became the fuel of choice as well as the preferred feedstock for the chemical industry. The availability of relatively inexpensive oil, due to the low cost of transportation by large tankers and pipelines, led to the dominance of oil. Another low-cost hydrocarbon feedstock is natural gas. The distribution of feedstocks used for primary petrochemicals in the United States in 1981 was 60% petroleum, 39% natural gas, and 1% coal. The increased cost of finding oil and gas today has again focused attention on a shift from crude oil and natural gas to coal. The coal will primarily be used to produce syngas, a mixture of carbon monoxide and hydrogen, which is the basic feedstock for C_1 chemicals. Syngas is readily converted to methanol, a building block for many C_1 and C_2 chemicals. Methanol is also used as a gasoline additive, and it can be readily converted to gasoline by the Mobil process.

The conversion technology is always an important variable. Today we are witnessing the emergence of new processing technologies that will in time revolutionize the manufacture of industrial organic chemicals. In the twenty-first century industrial organic chemicals will be produced by emission-free biomimetic processes. In the most efficient chemical factory, the human body, chemical reactions take place at room temperature in an aqueous environment and at a neutral pH in very high yields and with high selectivities. In industrial chemical plants chemicals are reacted in hot pipes over solid catalysts to produce a host of undesired by-products. Biotechnological processes, such as microbiological fermentation based on renewable biomass, will gradually become more important in the production of high-value products, such as pharmaceuticals, veterinary medicines, and agrochemicals.

New organometallic catalysts are emerging that allow the fixation of oxygen and nitrogen from air. In the future, new direct oxidation and amination processes will eliminate today's coproduct processes.

We have so far synthesized or identified over 5 million organic chemicals. Of these, only 60,000 are of commercial importance and a handful constitute the feedstocks for today's chemical industry. They are the building blocks of all the major plastics, of modern synthetic drugs, and of the insecticides and herbicides needed to produce food for the growing world population. They are also the feedstocks for the dyestuffs required to color our visual world. Other important areas of applications include synthetic fibers and rubbers, detergents, food chemicals, perfumes, and flavors as well as fuels, explosives, and propellants.

This book will not review classical synthetic methods. It will simply teach how we are, chemically speaking, doing things today. The basic chemical processes used in today's chemical plants are described, and emerging technologies are emphasized.

Before World War I, Germany and England were the leaders in industrial organic chemistry. Then the United States assumed the position of leadership by building huge plants and thereby utilizing the economies of scale. Today the United States is hampered by a giant bureaucracy, and the very existence of the chemical industry is threatened by a litigious society. Asian countries, especially Japan, Taiwan, and Korea, are rapidly catching up, outperforming the U.S. chemical industry in many areas. Tomorrow some third world countries will assume leadership in the manufacture of basic organic chemicals. Saudi Arabia and Mexico will emerge as large-scale producers of industrial organic chemicals by building huge plants next to their oil fields. Eventually a coal-based chemical industry may emerge in the United States and in Europe if the many political obstacles associated with the mining and handling of coal can be overcome.

2. Raw Materials

Introduction

The availability of raw materials and the energy to convert them into finished products has at all times been of major importance to the chemical industry. Organic chemicals are derived from carbon, hydrogen, and water, and therefore coal, natural gas, and oil are the basic raw materials for the synthesis of a myriad of industrial organic chemicals. Natural gas and oil especially will have to play major roles as both chemical feedstocks and primary energy sources well into the next century. Because of its unique value as a chemical feedstock, petroleum should be reserved for the chemical industry, and alternative sources of energy, especially coal and nuclear energy, should be allocated for fuel and electricity.

In Table 1 the 1984 production of crude oil in the major oil-producing countries is shown, as well as the known available reserves. If oil production continues at today's rate, the U. S. reserves will be used up by 1996.

Table 1. World Oil Production and Reserves, 1984

Country	Annual Production (10^9 barrels)	Reserves (10^9 barrels)	Reserves and Production (years)
Kuwait	0.36	90.0	250
Iraq	0.43	44.5	104
Saudi Arabia	1.71	169.0	99
United Arab Emirates	0.44	31.9	73
Iran	0.80	48.5	61
Libya	0.40	21.1	52
Mexico	1.10	48.6	45
Venezuela	0.68	25.8	38
Nigeria	0.51	16.7	33
United Kingdom	0.94	13.6	14
USSR	4.53	63.0	14
United States	3.79	34.5	9
World Total	21.10	707.2	34

The oil crises of 1973 and 1979 and the formation of OPEC created a need to reassess energy options in the major industrial countries. In the United States in 1980 the production of domestic oil and natural gas still accounted for over 50% of the required energy. In contrast, Western Europe and Japan have to import most of their fuel and feedstocks.

Natural gas is abundant in the Eastern Block countries, in the near East, and in the United States and Canada, and polyolefins are already produced near the wellhead in Saudi Arabia. Also, methanol is being produced near the wellhead in

Canada. Liquefaction of natural gas and transportation in large LNG tankers has made wellhead gas available for heating in the industrial countries. Western Europe is served by a major pipeline from Russia. Current free world gas reserves total almost 56.6 trillion cubic meters. The geographical distribution of proven natural gas reserves is listed in Table 2.

Table 2. Geographical Distribution of Proven Natural Gas Reserves

	Total Reserves (%)
Comecon countries	38
Near East	24
United States and Canada	13
Africa	9
Western Europe	8

Coal

Coal is by far the most abundant source of fossil energy in the world, and therefore conversion of coal into chemical feedstocks will gain in prominence by the turn of the century. Coal gasification could be utilized in the production of gasoline via syngas and methanol. Gasification and methanation of coal is accomplished in large fluidized bed gasifiers at about 700°C. The catalysts used are aqueous solutions of potassium hydroxide or potassium carbonate. The chemistry is shown in Figure 1.

$$2\,C + 2\,H_2O \longrightarrow 2\,CO + 2\,H_2 \qquad \Delta H = 268KJ$$

$$CO + H_2O \longrightarrow CO_2 + H_2 \qquad \Delta H = -34KJ$$

$$CO + 3\,H_2 \longrightarrow CH_4 + CO_2 \qquad \Delta H = -226KJ$$

$$2\,C + 2\,H_2O \longrightarrow CH_4 + CO_2 \qquad \Delta H - 8KJ$$

Figure 1. Coal gasification and methanation.

Underground coal gasification technologies have also been demonstrated. These methods will be especially valuable in the exploitation of marginal coal reserves that would otherwise remain unrecoverable.

Prior to the growth of the petrochemical industry, coal was the major source of industrial organic chemicals. For example, in 1950 about 95% of the benzene produced in the United States was derived from coal. In 1978 only 3% came from coal. All coal-derived chemicals are by-products of coke production for the iron and steel industry. In this process basically two types of chemicals are generated: light oil and coal tar. The major chemicals derived from coal are listed in Table 3.

The coal tar fraction also contains anthracene, naphthols, cresols, and pyridines. Large quantities of coal tar and some light oil are burned for fuel rather than being processed for chemical use.

Table 3. Major Chemicals Derived from Coal in the United States, 1981

	Production (tons/yr)
Light Oil	
Benzene	334,000
Toluene	63,000
Xylenes	13,000
Coal Tar	
Naphthalene	159,000
Coumarone-indene resins	32,000
Phenol	4,000

The conversion of coal into liquid fuels was of great interest to countries with no natural gas or oil sources. Pioneering work by F. Bergius in Germany prior to World War I led to an efficient destructive hydrogenation process. At the end of World War II, 12 plants were operational in Germany, producing about 100,000 barrels/day of liquid aviation and motor fuel. A parallel effort by Franz Fischer and Hans Tropsch produced hydrocarbons from syngas, obtained by gasification of coal with steam and oxygen. At the end of World War II, nine Fischer-Tropsch plants produced about 15,000 barrels/day of highly paraffinic products suitable for diesel fuel and wax production. After World War II all plants in Germany were shut down or converted to other uses. South Africa started Fischer-Tropsch plants in Sasolburg in 1955, 1980, and 1982. Pure syngas is fed to the Fischer-Tropsch reactors, and methane is sold as pipeline gas or is reformed to syngas and recycled. The production of higher boiling hydrocarbons was optimized for liquid fuel consumption.

The oldest plant (Sasol I) uses a fixed-bed reactor in which the syngas is fed into 2000 tubes (12 m × 50 mm) containing an iron oxide catalyst. The tubes are

Table 4. Product Selectivity of Sasol Synthetic Fuel

Products	Fixed Bed (220°C) (% based on C Atom)	Fluidized Bed (325°C) (% based on C Atom)
Methane	2.0	10
Ethylene	0.1	4
Ethane	1.8	4
Propylene	2.7	12
Propane	1.7	2
Butylenes	3.1	9
Butanes	1.9	2
C_5-C_{11} (gasoline)	18.0	40
$C_{12}-C_{18}$ (diesel fuel)	14	7
$C_{19}-C_{23}$	7.0	–
$C_{24}-C_{35}$ (medium wax)	20.0	4
C_{35} (hard wax)	25.0	–

surrounded by boiling water. The operating temperature is 500 K and the pressure is 2.7 MPa. Sasol I produces waxes, liquid fuels, pipeline gas, and chemicals. The two newer plants (Sasol II and III) use fluidized bed reactors, in which primarily ethylene, gasoline, and diesel fuels are produced. The product distributions in the Sasol plants are shown in Table 4.

The reaction conditions can be controlled to produce mainly pipeline gas by using higher temperatures in the fluidized bed reactor or to produce mainly liquid fuels. Two catalytic downstream processes are needed to optimize liquid fuel production: hydrocracking of the paraffins (wax) and oligomerization of the C_3-C_4 olefins. In this manner, a 73% yield of diesel fuel can be achieved. The Mobil syngas (methanol reforming) process has advantages over the Fischer-Tropsch process, because up to 81% gasoline can be produced using a zeolite catalyst.

Natural Gas

In the United States, natural gas has been the feedstock of choice for primary petrochemicals. The constituents of natural gas are low-molecular-weight saturated hydrocarbons in the C_1-C_4 range. However, methane could be as much as 97% of dry natural gas. Louisiana and Texas account for nearly two-thirds of the natural gas produced in the United States. The natural gas liquids (C_2-C_4 components) are used as feedstock for olefin production. These liquids are about 5–10% of the total, and the ethane/propane/butane ratio is about 4:2:1. In 1970 about 90% of the olefins produced came from natural gas, but the amount changed to 43% by 1978 because declining natural gas supplies prompted a switch to liquid feedstocks.

The major chemical produced from natural gas liquids is ethylene. However, propylene, butenes, and butadiene are important by-products. The most efficient feedstock for ethylene is ethane, because higher boiling liquid feedstocks produce more coproducts.

The major industrial chemicals derived from natural gas are shown in Figure 2. Ethylene, propylene, and butadiene are important monomers for high-volume industrial polymers.

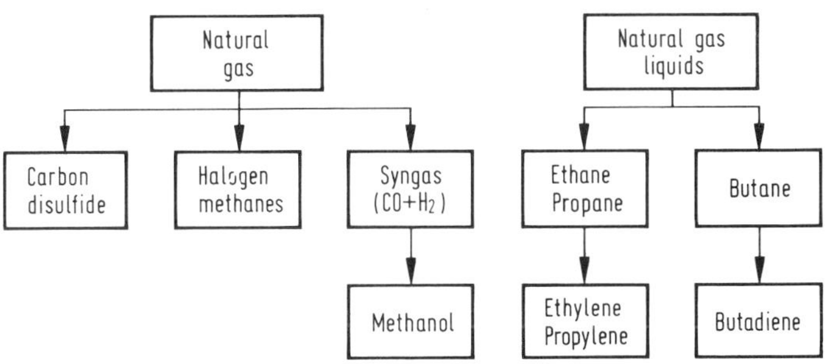

Figure 2. Major industrial chemicals based on natural gas.

In addition to the above industrial chemicals, hydrogen is produced from natural gas via syngas. Hydrogen is used as liquid rocket fuel, in the synthesis of ammonia, and in refining operations. The Columbia space shuttle uses 381,000 gallons of hydrogen (43.2 million cubic feet) each time it is launched into space. Catalytic reforming, steam reforming, and hydrocracking consume approximately 30% of the total amount of hydrogen produced in the United States every year. The more important chemical uses of hydrogen include the synthesis of methanol and oxo alcohols, the production of cyclohexane, and the manufacture of aniline, diaminotoluene, and hexamethylenediamine.

Crude Oil

Crude oil is the base for most of today's feedstocks for industrial organic chemicals, and of course it is the main source for motor gasoline and heating oil. The first step in oil-refining operations is the separation of the crude oil into fractions with different boiling points, which is achieved by fractional distillation. The typical fractions obtained in modern refineries are listed in Table 5.

Table 5. Typical Crude Oil Fractions

	Boiling Point (°C)
Light naphtha	20–100
Heavy naphtha	100–150
Kerosene	150–235
Light gas oil	235–343
Heavy gas oil	343–565

About 50% of the gasoline sold in the United States is obtained from petroleum by catalytic cracking. A fluidized bed process is usually used, and the purpose of catalytic cracking is to convert heavy distillate to compounds in the boiling range of gasoline and middle distillate. The temperatures used in this cracking process range from 480 to 570°C, and the catalyst is usually silica gel activated with 15–60% Al_2O_3. The new, more active, zeolite catalysts operate at higher temperatures.

The fractions boiling from 80 to 230°C have too low an octane rating to be useful as gasoline. Catalytic reforming is used to upgrade the so-called naphtha materials. This process is conducted at 430–520°C in the presence of platinum on alumina. The platinum level is on the order of 0.5%, and chlorine is sometimes added as a promoter. The hydrogen that is generated is recycled to minimize formation of carbonaceous deposits. The distillate is pretreated to remove sulfur, nitrogen, arsenic and lead impurities that would deactivate the catalyst. This pretreatment is usually done by hydroprocessing, a treatment with hydrogen that also converts olefins to alkanes. A fixed-bed process is often used. The catalysts are either cobalt/molybdenum or nickel/tungsten, and the reaction is conducted at 350–500°C.

In the catalytic reforming process, cycloparaffins are converted into aromatic molecules and linear molecules are converted into branched and, to a lesser degree, cyclic compounds.

Alkylation is another process used to increase the octane rating in the gasoline pool. In this reaction the lower olefins (C_3–C_5) are reacted with isobutane in the presence of sulfuric or hydrofluoric acid to give higher molecular weight compounds.

It is expected that in the future unleaded gasoline will be composed of mostly catalytic cracker gasoline, straight-run gasoline, and some reformate. Unleaded premium gasoline will be composed mainly of alkylate and reformates with straight-run gasolines and butanes added.

The distribution of fuel products manufactured in the United States in 1980 is given in Table 6.

Table 6. Distribution of U. S. Petroleum Products, (1980)

	Total Production (%)
Gasoline	43
Diesel and heating oil	20
Jet fuel	9
Residual fuel oil	9
Asphalt and coke	6
Gas	4
Petrochemical feed	2
Others	7

Other processes used in an oil refinery include hydrocracking, the conversion of higher molecular weight compounds to lower-boiling materials, and hydroprocessing.

In hydroprocessing olefins are converted to paraffins, wax is removed from lube stock, and unstable by-products obtained in steam cracking are stabilized. The use of hydroprocessing in the removal of impurities was discussed earlier.

In addition to fuel products, nearly all refineries produce raw materials for the chemical industry. Light olefins are made by thermal cracking in the presence of steam. Thermal cracking is conducted at 800–850°C and slightly above atmospheric pressure. This process is important because olefins are not present in crude oil. Catalytic cracking and hydrocracking are also used for the manufacture of olefins, and catalytic reforming is used to obtain aromatic molecules (BTX fraction). Ethane, butane, and naphtha are used as feedstocks for steam cracking, while naphtha and gas oil fractions are used for catalytic reforming. The availability and price of the basic oil-derived organic chemicals depend entirely on the consumption of gasoline. In recent years aromatic hydrocarbons became expensive because the governmental requirements for lower lead content in gasoline forced refiners to utilize the BTX (benzene, toluene, xylene) fraction as octane boosters in gasoline.

The major petroleum-based industrial organic chemicals are shown in Figure 3.

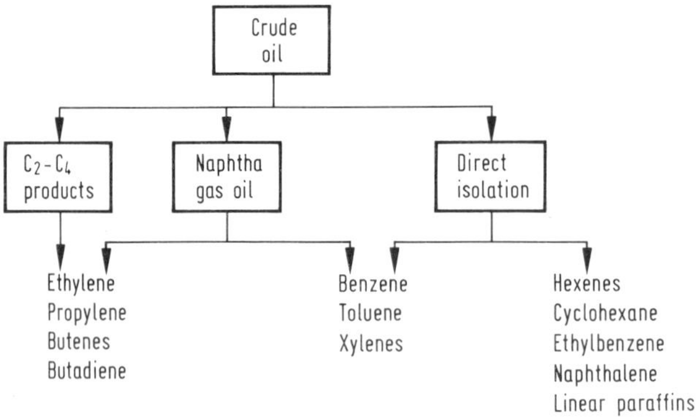

Figure 3. Petroleum-based major industrial organic chemicals.

In oil- and gas-producing regions in the world, where natural gas liquids used to be flared (Middle East, Mexico, North Sea), new olefin plants based on this feedstock are under construction or are in the planning stage. In Canada, huge methanol plants have been constructed that use syngas produced from readily obtainable natural gas feedstocks.

The world's largest primary petrochemical producers are listed in Table 7.

Table 7. Worldwide Petrochemical Producers, 1983

	Capacity (1000 tons/yr)
Shell Group	8.450
Exxon	8.240
Dow	6.040
State complex (Russia)	5.680
DuPont	4.180
Arco	3.970
ICI	3.930
Standard Oil (Indiana)	3.410
UCC	3.370
Phillips	3.300
BP	3.110
Mitsubishi Group	2.900
Pemex (Mexican Govt.)	2.640
ENI (Italian Govt.)	2.620
Texaco	2.460
Montedison	2.190

Biomass

A potential alternative feedstock for the chemical industry is biomass. The attraction of biomass as chemical feedstock is its virtual inexhaustibility. However, in reality, fertilizers, nutrients, and insecticides are needed to form biomass. The main polysaccharide in plants is cellulose, which is produced and destroyed in nature at the rate of 100 billion tons of carbon each year. Cellulose products are used directly as fuel or contruction material (wood), or in form of fibers (cotton, hemp, sisal, linen). Other traditional uses of cellulose include the manufacture of rayon, cellulose acetate, and cellulose ethers.

Waste cellulose, which does not require additional farmland, is an abundantly available raw material. Aerobic conversion of cellulose leads to glucose, whereas anaerobic conversion leads to hydrocarbons (chiefly methane). Industrial fermentation of natural products produces ethanol, acetone/butanol, citric acid, gluconic acid, and itaconic acid. Lactic acid is also made by an anaerobic fermentation method.

Sucrose is produced for food consumption from sugarcane or sugar beets. Sorbitol is manufactured by hydrogenation of glucose. Linseed, soybean, cottonseed, coconut, palm, and other plant oils and animal fats have always been important raw materials for the chemical industry. Castor oil is transformed into the basic building block for nylon-11. Corn cobs and oat hulls are used to produce furfural, which is used in resins and is also chemically converted into tetrahydrofuran (see p. 124). Wood-derived chemicals include dimethylsulfoxide and vanillin, and steam distillation of wood is used to produce turpentine oil, pine oil, and pinenes.

A number of processes are under development to gasify municipal waste, sewage sludge, and other forms of biomass by partial combustion. The generated low-Btu fuel gas (syngas) could be used in the synthesis of C_1 chemicals as well as in the manufacture of ammonia and in the oxo process. Anaerobic digestion could generate syngas from the same waste products.

Future Outlook

New processing technologies are emerging that will revolutionize the manufacturing of industrial organic chemicals. In the nineteenth century industrial organic chemicals were based on coal, in the twentieth century we entered the petrochemical era, and in the twenty-first century industrial organic chemicals will be manufactured by emission-free biomimetic processes.

In today's petrochemical world the basic feedstock – natural gas or crude oil – is reorganized to produce a mixture of hydrocarbons designed to be used as motor fuel. Computerized controls are used to optimize the reaction conditions to achieve the desired product mix. The petrochemicals are basically by-products in the production of gasoline. Large chemical companies are merging with oil companies to secure the price and availability of the chemical feedstocks. Coal is an alternative source of hydrocarbons, but it will not be important as a chemical feedstock in the forseeable future.

Biogenetic engineering will play a role in changing our chemical world. Biomass may be more readily adaptable to farming because advanced green plants can be developed that fix nitrogen from the air, eliminating the need of some fertilizers. Also, fermentation processes could be vastly improved by genetic engineering. These developments would benefit the production of complex natural products, such as antibiotics and vaccines. However, for most of our basic chemical feedstocks it is necessary to improve yields and selectivities by using more sophisticated catalysis and eventually establish an industry that stands on its own feet and is no longer dependent on the unpredictability of gasoline consumption.

The natural distribution of elements on earth assures an ample supply of oxygen, the most abundant of the elements (49.2%), hydrogen (0.87%), and carbon (0.08%), the building blocks of industrial chemicals.

3. Production of Basic Industrial Chemicals

Introduction

The petrochemical industry has almost eliminated the use of coal and coal tar as sources of chemical raw materials. It has also replaced many of the traditional methods of manufacture of industrial organic chemicals derived from natural products. Modern plastics and synthetic fiber and rubber products are based almost exclusively on petrochemical feedstocks. Chemical production complexes are usually located in the vicinity of oil refineries, and many of the chemicals needed in further transformations of the basic olefins and aromatic feedstocks are supplied via pipeline. A typical modern chemical complex is the Houston area in Texas, where numerous oil refineries and chemical plants are located along the shipping channel.

Trends in the production of basic industrial organic chemicals favor location of chemical plants near natural gas and oil fields. Natural gas liquids are collected near the wellhead and piped to strategically located gas plants. There the gas is treated to remove sulfur compounds and carbon dioxide. Methane is compressed and distributed for use as fuel, and the natural gas liquids are used as feedstock for olefin production in petrochemical complexes near the oil fields. Typical examples are the chemical complexes at Yanbu and Jubail in Saudi Arabia. Although coal has a lower cost than most liquid fuels, it is not competitive, because the handling and processing of liquids is considerably less expensive. An exception is the Sasol complex in South Africa, where attempts are under way to reestablish a chemical industry based on coal because of political considerations.

Table 8. Production of Primary Industrial Organic Chemicals in the United States, 1984 (1000 metric tons)

Ethylene	12,993	Acetic acid	1,270
Propylene	6,315	Phenol	1,187
Toluene	6,144	Butadiene	1,050
Benzene	5,549	Acrylonitrile	975
Ethylene dichloride	5,112	Formaldehyde	907
Ethylbenzene	3,571	Vinyl acetate	891
Styrene	3,175	Propylene oxide	881
		Acetone	851
Vinyl chloride	3,159	Cyclohexane	767
Methanol	3,010	Isopropanol	550
Terephthalic acid and		Caprolactam	445
dimethyl ester	2,585	Phthalic anhydride	386
Ethylene oxide	2,534	Aniline	302
Ethylene glycol	2,027	Bisphenol A	292
p-Xylene	1,867	Maleic anhydride	136
Cumene	1,501		

The current U. S. production of industrial organic chemicals is approximately 95 million metric tons, 99% of which is accounted for by the top 100 chemicals. Of these, approximately 74% are produced from five primary feedstocks: ethylene, propylene, benzene, toluene, and xylene. Table 8 liste the primary industrial organic chemicals produced in the United States in 1984. Many of these chemicals are monomers or intermediates in the production of industrial polymers.

Today's chemicals are produced by simple processes, such as direct oxidation, ammoxidation, and dehydrogenation. Therefore, petrochemicals are often made from hydrocarbon raw materials with the same number of carbon atoms as the finished products. The addition of one carbon is accomplished by hydroformylation, using syngas, a mixture of carbon monoxide and hydrogen, to accomplish the chain extension (oxo process).

Chemical Operations

Chemical plants usually have reactor sections in which the chemical transformations are accomplished and finishing sections where the mixtures of chemical products are separated. Since cooling water is required, chemical plants are often located near water, and water is also used to remove and dilute chemical waste products. A typical example is in the Rhein-Main area in Germany, where BASF, Hoechst, and Bayer have huge chemical plants located on the riverbanks. When oil refineries are established near the coast (Houston, Rotterdam), chemical complexes are also built near the coast. This is especially valid for Japan, where all raw materials are imported by sea.

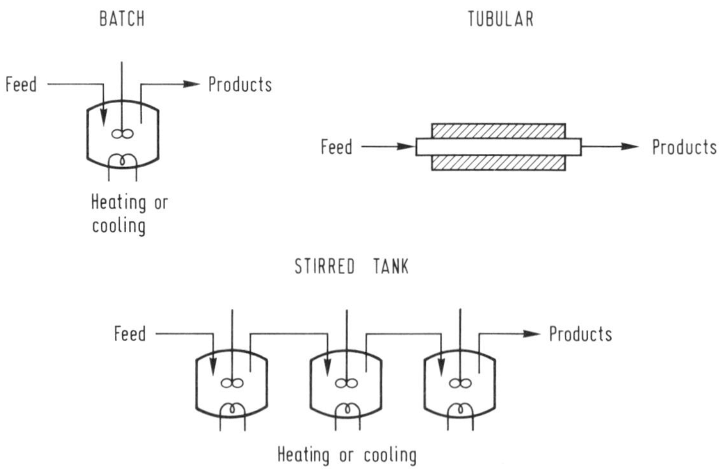

Figure 4. Basic reactor types.

Chemical transformations are conducted in reactors in either batch or continuous operations (see Figure 4). A batch reactor is usually a tank with or without agitation, and jacketing provides heating and cooling. The heat exchangers can be located inside or outside of the tank. Batch reactions are used primarily in small-scale production. In continuous reactors the reactants are introduced and the products are withdrawn simultaneously. Continuous operations facilitate control of product quality and reduce operating cost. The reactor configurations include tanks, tubular reactors, and towers. A combination of batch and continuous operation is called a semicontinuous process. Because back-mixing is achievable in stirred tank reactors, many slow reactions are conducted using a series of tank reactors. Tubular reactors are used when gas-liquid reactions are conducted on solid catalysts. A single continuous tube or several tubes in parallel are used, and heat transfer to or from the reactors can be achieved with jacketing or the use of a shell tube design. The tubular reactors are packed with catalyst pellets or inert solids (fixed-bed reactors).

Vertical cylindrical or tubular vessels containing fine solid particles are called fluidized bed reactors. The solid particles can be either reactants or catalyst. The fluid stream of reactants is introduced at the bottom of the reactor at such a rate that the solids are floated in the fluid stream without being carried out of the reactor. A typical example is the fluidized bed reactor used in the Synthol process operational in the Sasol II Fischer-Tropsch plants located in Secunda, South Africa (Figure 5).

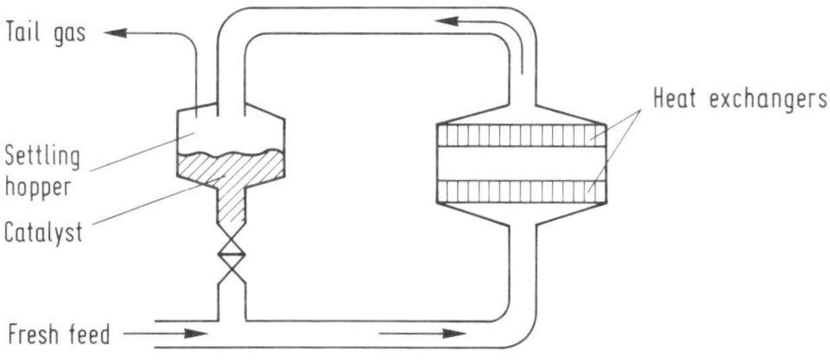

Figure 5. Fluidized bed reactor used in the Fischer-Tropsch production of liquid fuels from coal.

Feed gas enters the bottom and meets a stream of downflowing hot catalyst from the standpipe, which heats the gas to the ignition temperature. The gas passes through two banks of heat exchangers to the settling hopper above the standpipe, where the gas and catalyst disengage. Gas and hydrocarbon product vapors leave the reactor via cyclones located at the settling hopper.

Slurry phase reactors are vertical columns containing fine catalyst particles slurried with a liquid medium that may be one of the reactants. Such reactors are used in the hydrocracking of residual fuel oils. This process converts feedstocks ranging from naphtha to high-boiling residues into more valuable lower-boiling products. Hydrocracking catalysts have a dual function; they consist of metal sites on an acidic cracking base.

Tower reactors are vertical cylindrical structures containing baffles or solid packing that are often employed in continuous processes involving heterogeneous reactions. The solid packing could consist of a catalyst or simply an inert substance. Continuous distillation columns operate similarly. Usually the feed is introduced at one or more points along the column shell. Trays or plates are used for separation of the liquid phase, which runs down the column while the vapor goes up and is condensed at the top of the column. The overall separation achieved between the overhead and bottom products depends primarily on the vapor pressures of the components. Design variables include stream concentrations, temperatures, pressure, and flow rates. Extractive distillation, used in the separation of close-boiling liquids, uses a third component solvent. However, this method is used only when fractional distillation is not feasible, because energy is required to boil and recycle the third component.

Separation of reaction products by distillation is the preferred method, but crystallization and liquid extraction processes are also very common. Crystallization is very important in the isolation of solid products that have a very low vapor pressure and for thermally labile products. Crystallization usually affords highly purified solids from relatively impure solutions in a single processing step. Filtration or centrifugation is used to separate crystallized solids from the mother liquor. Liquid extraction requires reboiling of the solvent.

Wacker Process. An Example

A typical modern industrial process is the Wacker process for the production of acetaldehyde by direct oxidation of ethylene (Figure 6). The development of this process was prompted by the availability of inexpensive ethylene from the cracking of natural gas and naphtha. The classical process of manufacture of acetaldehyde is based on the hydration of acetylene. The Wacker process, developed in 1957–1959 by Wacker and Hoechst, consists of oxidizing ethylene to acetaldehyde using a cupric chloride–palladium chloride catalyst system. There is also a two-stage version in which the spent catalyst is regenerated in a separate reactor; this process uses air instead of oxygen. Material of construction is titanium, and the reactor is rubber-lined and further protected by an acid-resistant brick wall.

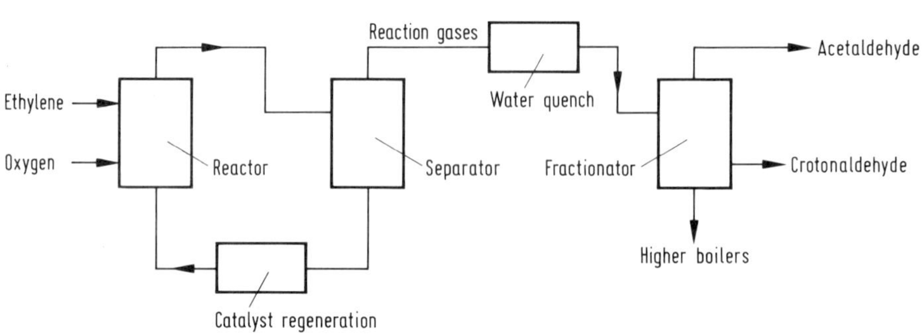

Figure 6. The Wacker process.

In the single-step process, oxygen and ethylene are fed into the catalyst solution at 3 bar and 120–130°C. The conversion of ethylene is on the order of 40%. In the two-step process, ethylene is almost completely converted. After reducing the pressure, acetaldehyde/water mixtures are distilled, and the catalyst solution is regenerated using air at 100°C and 10 bar. The advantage of the total ethylene conversion is balanced by the need to use an additional reactor (Figure 6).

The Wacker process is a typical example of the use of catalysts in modern industrial chemical processes. Platinum chloride is the actual catalyst, and the cocatalyst (cupric chloride) reoxidizes the zero-valent palladium back to the divalent state. This catalyst system is also used in the oxidation of propylene to acetone.

Catalysis

Basically two types of catalytic processes are used in the chemical industry: homogeneous catalysis, in which the catalyst is dissolved in the reaction mixture, and heterogeneous catalysis, in which the catalyst is deposited on an inert solid surface, usually silica or aluminum oxide. The Wacker process is a typical example of homogenous catalysis.

Regeneration of the catalyst *in situ* is the preferred method, and the copper redox systems used in hydroxylation and acetoxylation processes are typical examples. A recent development is the carbonylation of methanol to acetic acid. Rhodium carbonyl catalysts, which are also used in hydroformylation processes, undergo an oxidative addition to methyl iodide, which is the cocatalyst. The total reaction sequence is shown in Scheme 1.

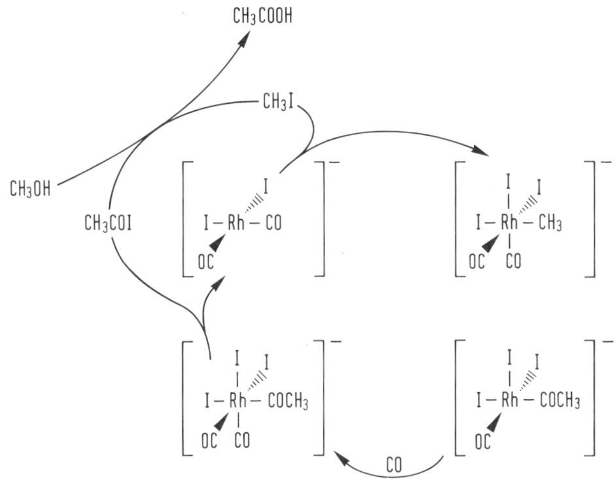

Scheme 1. Carbonylation of methanol to acetic acid.

Important industrial processes using homogeneous catalysis are listed in Table 9. Homogeneous catalysis is more useful in processes that do not require isolation of the catalyst from reaction residues. For example, the direct conversion of dinitrotoluene with carbon monoxide to toluenediisocyanate is feasible in the laboratory but is not used in industrial plants because of losses of the noble metal catalyst on isolation from the distillation residue.

Table 9. Industrial Processes Using Homogeneous Catalysis

Process	Catalysts	Examples
Oxidation	Salts of Co, Ni, Cr, Mn, Cu	Cyclohexane → Cyclohexanone + cyclohexanol
		Benzene → maleic anhydride Xylenes → dicarboxylic acids Cumene → phenol + acetone Naphthalene → phthalic anhydride
Hydroxylation	$PdCl_2 + CuCl_2$	Ethylene → acetaldehyde Propylene → acetone
Acetoxylation	$PdCl_2 + CuCl_2$	Ethylene + acetic acid → vinyl acetate
		Butadiene + acetic acid → 1,4-butanediol
Hydroformylation	CO, Rh	Propylene → butyraldehydes
Carbonylation	Rh carbonyls + CH_3I	Methanol → acetic acid
Olefin metathesis[a]	Mo, W carbonyls	Propylene → ethylene + butene
Polymerization	$TiCl_3$, Et_2AlCl[b]	Ethylene → polyethylene Propylene → polypropylene

[a] Disproportionation.
[b] Ziegler-Natta catalyst.

The basic concept of heterogeneous catalysis involves transitory adsorption of one or more of the reactants onto the surface of the catalyst, rearrangement of the bonding, and desorption of the products. The temperature is usually between about 20°C and 500°C. However, several important processes (oxidation of methanol to formaldehyde, steam reforming of natural gas and naphtha, etc.) are carried out at much higher temperatures.

The solid support for the heterogeneous catalysts is usually silica, alumina, or a natural or synthetic (zeolite) clay. Some catalysts have more than one function. For example, metal sites can be attached to an acid clay, redox function can be built into the metal centers using metals with more than one stable oxidation state, and dehydrogenation sites can be incorporated into the acid catalysts used in thermal cracking.

Industrial catalysts generally fall into one of several major categories, including acids and bases, metals, metal oxides, and metal sulfides. Hydrogenation catalysts are transition metals and isomerization catalysts are strong acids. Coke formation is a problem with most of the hydrocarbon catalysts. The coke formation can be reduced using a second metal along with the typical platinum catalyst. Rhenium and iridium are good examples.

In recent years so-called shape-selective catalysts, such as zeolites (aluminosilicates) and aluminophosphates, have been used to achieve better selectivity, as in the p-alkylation of toluene with ethylene using a zeolite catalyst and the selective dehydration of butanol on a Linde Type 5A zeolite. Isobutyl alcohol cannot enter the catalyst pores. Also, n-paraffins can be selectively hydrocracked in Mobil's Selecto-forming process. Branched or cyclic paraffins are not cracked, because they are unable to enter the pores of the catalyst. Zeolites are also used as absorbents and as detergent builders. Table 10 summarizes the commercial use of zeolite catalysts.

Table 10. Commercial Use of Zeolite Catalysts

Process	Zeolite Type	Reactor	Status
Catalytic cracking of petroleum	Rare earth hydrogen Y matrix	Fluidized bed	Widely used
Catalytic dewaxing	ZSM-5	Fixed bed	Several plants
Xylene isomerization	ZSM-5	Fixed bed	60% of world capacity
Ethylbenzene production	ZSM-5	Fixed bed	Three plants
Methanol to gasoline	HZSM-5	Fixed bed	One plant
Methanol to olefins	HZSM-5	Fluidized bed	Pilot plant

Most zeolites are synthetic materials because of the strict requirements for high purity, but some naturally occurring zeolites are used as industrial catalysts.

Some of the more important heterogeneous catalysts used in industry are listed in Table 11.

Table 11. Industrial Processes Using Heterogeneous Catalysis

Process	Catalyst	Example
Dehydrogenation	Chromia-alumina, Fe_2O_3, Cr_2O_3 ZnO	Butane → butenes + butadiene, Ethylbenzene → styrene Isopropanol → acetone
Hydrogenation	Ni, Pd	Benzene → cyclohexane
Oxidation	Ag Ag or Fe_2O_3/MoO_3 V_2O_5	Ethylene → ethylene oxide Methanol → formaldehyde o-Xylene → phthalic anhydride, Butane → maleic anhydride
Ammoxidation	Complex metal molybdates	Propylene → acrylonitrile
Acid-catalyzed reactions	Zeolite	Catalytic cracking
	Pd/zeolite	Hydrocracking
	Pt/alumina or zeolite	Isomerization
	Pt or Pt/Re/alumina	Catalytic reforming
	ZSM-5 zeolite	Methanol → gasoline
Syngas reactions	Fe_2O_3 Ni ZnO/Cr_2O_3	Fischer-Tropsch synthesis $CO + 3H_2 \rightarrow CH_4 + H_2O$ $CO + 2H_2 \rightarrow CH_3OH$

Waste Disposal

Even in the best industrial processes, yields are far from quantitative. A typical multiprocess industrial chemical plant uses raw materials, water, oxygen, and nitrogen to produce about 80% chemicals, the remaining 20% being solid, liquid, and gaseous waste products. Some of the inorganic and organic waste products contaminate the waste water from such a plant. In contrast, cooling water remains clean. For example, Hoechst in Germany uses approximately 5% of the water from the Main River to cool its plants (about 630,000 cubic meters per day) and the temperature of the river water is increased by only 0.6 K. Some processes, such as the manufacture of organic dyestuffs, produce considerable amounts of by-products, while others, for example, the manufacture of polyethylene, are almost emission-free.

Modern process research focuses on the development of new emission-free and energy-efficient processes, because the cost of managing the removal of waste products is increasing. Waste disposal in the United States is regulated under the Resource Conservation and Recovery Act (RCRA). Many chemical waste products are labeled hazardous, and the generator of waste products is responsible for their disposal, which includes treatment, storage, disposal, and recycling. In 1987 a total of 205 million metric tons of hazardous waste was treated in the United States, 99 million metric tons was stored, and 32 million metric tons was disposed of in approved facilities. In addition, 182 million metric tons of RCRA-exempt waste materials was treated, and 43 million metric tons was recycled.

Elimination of coproduct processes is another research objective, because the coproduct may have a declining market potential. A typical example is the manufacture of phenol by the oxidation of cumene. The coproduct acetone is being used less then formely in its two major outlets, as a reactant in methyl methacrylate production and as a solvent. New improved processes for the production of methyl methacrylate are on the drawing board, and volatile organic solvents are gradually being replaced by aqueous systems.

Recycling of by-products and coproducts will also become more pronounced because of the increasing cost of waste disposal. Conceivably acetone could be converted to propylene and water, and the propylene could be used to manufacture cumene. This indirect recycling scheme avoids the formation of a coproduct. Economic factors determine the viability of chemical processes, and chemical ingenuity will be able to cope with the new requirements. Waste treatment, disposal, and recycling also require chemical transformations. In contrast, storage of chemical waste will become unacceptable.

Chapter II. C_1 Chemicals

1. Introduction

The basic chemical feedstocks for one-carbon chemistry are methane and synthesis gas. Methane is obtained from three sources: natural gas, refinery gas, and gas produced at the wellhead casings. Previously, wellhead gases and refinery gases were either used as boiler fuel or flared. Today huge LNG tankers transport liquefied natural gas from wellhead areas to other parts of the world where it is needed. Another method of utilization of wellhead gas is its conversion to methanol via syngas because methanol is easier to transport.

Methane is a relatively unreactive molecule. However, both homogeneous and heterogeneous activation are feasible, and the addition of oxidants is helpful in many reactions. Ziegler-Natta catalysts activate methane, and if the reaction is conducted in the presence of ethylene or acetylene the expected C_3 derivatives, propane and propene, can be obtained. Also, a radical-type activation, using chlorine to generate radicals as well as the oxidizing agent, results in formation of ethane and ethylene from methane. This reaction is conducted at high temperatures (700–1700°C) and for very short reaction times.

$$CH_4 + Cl_2 \longrightarrow C_2H_6 + C_2H_4 + HCl + H_2$$

Of course, chlorination of methane is an expected side reaction. Another problem related to this type of reaction is the choice of reactor material because of corrosion.

Reaction of methane in the presence of oxygen, using platinum, zeolite, or a variety of metal oxides (antimony, bismuth, tin, lead, thallium, manganese, cadmium, titanium, zinc, and silicon) can lead directly to methanol, formaldehyde, ethane, and ethylene. In view of the availability of methane, it is expected that processes such as the oxidative coupling of methane to ethane and ethylene, the direct formation of methanol and formaldehyde, and the oxidative methylation of acetonitrile to acrylonitrile will become important by the end of this century.

Important industrial chemicals obtainable from methane are listed in Table 12.

Table 12. Products Derived from Methane

Coreagent	Product
Water	Syngas
Ammonia	Hydrogen cyanide
Chlorine	Chloromethanes
Sulfur	Carbon disulfide
Oxidizing agents[a]	Ethane, ethylene
Acetonitrile[b]	Acrylonitrile
Ethylene or Acetylene	Propane, propene

[a] Oxidizing agents are chlorine or oxygen.
[b] Oxygen is also used to effect the oxidative methylation.

Synthesis gas (syngas) is a mixture of carbon monoxide and hydrogen in various proportions. The carbon monoxide/hydrogen gas combinations are also referred to as water gas, crack gas, or intermediate-Btu gas. Syngas is presently produced mainly from methane-rich natural gas streams, but it can also be obtained from the burning of oil, gasoline, coal, and biomass. Coal gasification is conducted with steam and oxygen, and the production of syngas from natural gas or oil in the presence of steam is similar. In the ICI reforming process, naphtha feedstock is used to produce syngas. The catalytic reforming is conducted at 700–830°C and 15–40 bar, using a $Ni/K_2O/Al_2O_3$ catalyst. A partial oxidation process, using all hydrocarbons from methane to crude oil residues, was developed by Shell. In the absence of a catalyst, the preheated feeds are reacted at 1200–1500°C and 30–80 bar with less than the stoichiometric amounts of oxygen and water. Impurities are removed by a variety of methods. The ratio of carbon monoxide to hydrogen can be adjusted using the so-called shift reaction in which carbon monoxide is converted by means of water into carbon dioxide and hydrogen. The shift catalyst consists of iron/chromium oxide mixtures and it is used at 350–400°C.

Syngas applications involve the production of methanol, aldehydes, and alcohols from olefins via hydroformylation reactions and the production of hydrocarbons using the Fischer-Tropsch process. The latter process is used by Sasol in South Africa.

In a variation of the Fischer-Tropsch process, Shell uses a two-step process to manufacture middle distillate. In the first step syngas is converted into a polymethylene wax (C_1–C_{50}) using a Fischer-Tropsch type of catalyst in a simple fixed-bed operation. The second step consists of hydroisomerization and hydrocracking to produce a mixture of naphtha, kerosene, and gas oil. This reaction is conducted in a trickle-flow reactor, and the output of kerosene and gas oil is optimized by operating the second step at low conversion per pass. Methanol can be produced in high yield and selectivity, but changes in the catalyst can afford blends of C_1–C_5 alcohols that can be used directly as octane enhancers and cosolvents. Attempts are under way to produce ethylene glycol from syngas, either directly or via methanol or formaldehyde. Syngas also serves as a source for hydrogen and carbon monoxide. For example, hydrolysis of carbon monoxide in the presence of sodium hydroxide at 115–150°C and 8–30 bar produces formic acid. Formic acid is used to produce *N,N*-dimethylformamide, a polar solvent for polymerization reactions.

While carbon monoxide is of importance in the manufacture of industrial organic chemicals, carbon dioxide has found only limited use. The largest volume use of carbon dioxide is in the synthesis of urea. Other uses include the Kolbe-Schmitt synthesis of salicyclic acid and the manufacture of ethylene carbonate from ethylene oxide and carbon dioxide.

Commercially important C_1 products include the oxygen-containing products methanol, formaldehyde, and phosgene; the nitrogen-containing products urea, hydrogen cyanide, and methylamine; and a large number of halogenated methane derivatives. The largest volume products are listed in Table 13.

Table 13. Commercially Important C_1 Chemicals

Product	U.S. Consumption, 1984 (1000 tons)
Urea	7148
Methanol	3763
Formaldehyde[a]	2485
Phosgene	580
Carbon tetrachloride	323
Methylene chloride	250
Methyl chloride	175
Chloroform	164

[a] 37% solution.

Several of the important C_1 chemicals are intermediates or monomers for industrial polymers. Urea is used in the formulation of urea/formaldehyde resins, and formaldehyde is also used in the manufacture of phenol/formaldehyde and melamine resins. A total of 1,7 million tons of formaldehyde-derived thermosetting resins was produced in the United States in 1984. Methanol is used in the synthesis of dimethyl terephthalate, the basic building block for polyesters, and phosgene is used in the manufacture of the commercial isocyanate building blocks for polyurethanes. Phosgene is also a comonomer for polycarbonates. Recently, attempts have been made to produce carbonate polyols from ethylene carbonate.

Syngas or methane-derived processes for C_2 chemicals (ethylene, ethylene glycol, and acrylonitrile) are under intensive development, and by the end of this century coal-based C_1 chemistry could play an important role in the production of industrial polymers. Today, only Tennessee Eastman uses coal-derived syngas in the production of acetic anhydride, which is used in the production of cellulose acetate.

2. Methanol CH_3OH

Consumption (1984):	United States	3,763,000 tons
	Western Europe	3,360,000 tons
	Japan	1,560,000 tons
	World (total)	12,000,000 tons
Production Process:	Reaction of carbon monoxide with hydrogen	
Major Uses:	Plastics and fibers (50%), formaldehyde (30%), acetic acid (10%), solvent (10%), fuel additive	
Price:	$ 0.1/liter	

Methanol is one of the most important industrial chemicals. Its use will increase in significance in the years to come, because methanol is the key intermediate in the emerging C_1 chemistry. The first large-scale methanol plant based on the reaction of carbon monoxide with hydrogen was built by BASF in Germany in 1923. The catalyst systems used in the high-temperature, high-pressure process (320–380°C, 340 bar) were modified oxide catalysts, such as ZnO/Cr_2O_3 or CuO/ZnO. The stoichiometry of H_2/CO in the synthesis gas used was adjusted by the addition of CO_2.

$$CO + 2\,H_2 \longrightarrow CH_3OH$$

Today the ICI low-pressure process (240–260°C, 50–100 bar), using a $CuO/ZnO/Al_2O_3$ catalyst, is most widely used in the manufacture of methanol. A similar process was developed by Lurgi (230–250°C, 40–50 bar) using a modified CuO/ZnO catalyst. Typical plant sizes are on the order of 1200–2500 tons/day. The carbon monoxide and hydrogen used in the manufacture of methanol is presently derived from oil and natural gas. However, coal-based syngas will become more important if oil and gas prices escalate. Also, technologies based on gasification of biomass are under consideration. Brazilian authorities are considering methanol from biomass as a supplement to their established ethanol fuel program. The biggest problem encountered in the conversion of biomass to methanol is the collection of the biomass and its variable character. In contrast, gasification of coal is well established. After successful operation of a pilot plant, Rheinbraun in West Germany is planning a commercial-scale lignite gasification plant using the Winkler process.

Methanol is currently used to produce formaldehyde, dimethyl terephthalate, methyl halides, methylamines, methyl methacrylate, and acetic acid. However, future fuel uses will accelerate the growth of methanol. World methanol demand by 1990 is projected to be 22.9 million tons, of which 3 million tons will be used in the production of MTBE (methyl tert-butyl ether) and 2.9 million tons will be used in low-level methanol/gasoline blends. Currently Arco in the United States markets Oxinol, a 1:1 mixture of methanol and tert-butanol, as a gasoline additive. This mixture is added to gasoline in amounts of up to 9 Vol-%.

The advantage of using methanol as a fuel additive is that it serves as an antiknock agent. If gasoline contains only 5% methanol, its lead content can be lowered from 0.45 to 0.15% without a change in knocking. An even more widely used approach for increasing the octane rating of unleaded gasoline involves the addition of MTBE, which is readily produced from methanol and isobutylene using an acidic clay catalyst.

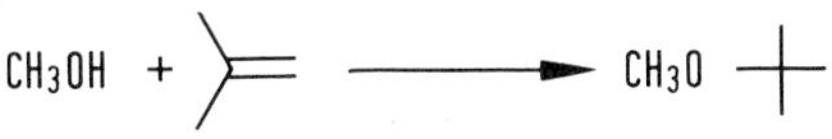

MTBE is lower in cost than methanol/tert-butanol mixtures. Another attractive approach to the production of gasoline additives uses molybdenum sulfide as the catalyst in the syngas reaction. This catalyst allows the manufacture of a variable blend of C_1-C_5 alcohols.

A new process developed by Mobil achieves homologation of methanol to produce gasoline. This process uses a highly specific synthetic zeolite (ZSM-5). The first step in this reaction, dehydration of methanol to dimethyl ether, is followed by further dehydration to yield olefins. These olefins are rearranged on the catalyst to give a gasoline-type mixture consisting of alkanes and aromatic hydrocarbons. The first commercial installation of the Mobil MTG (methanol to gasoline) process in New Zealand in 1986 uses natural gas as feedstock. A fluidized bed version of this process is under investigation. In this version, crude methanol is vaporized and superheated to 180°C before being sent to the reactor. Pressure in the reactor is less than 60 psig, and the temperature range is from 370 to 430°C. Another alternative for the production of motor fuel is the homologation of methanol to ethanol, which is a better gasoline blending compound. The reductive carbonylation reaction proceeds as follows:

$$CH_3OH + 2\,CO + H_2 \longrightarrow CH_3CH_2OH + CO_2$$

The catalyst system used is a mixture of iron carbonyl compounds and tertiary amines.

The fastest growing chemical application of methanol is in the manufacture of acetic acid based on the low-pressure carbonylation route pioneered by Monsanto. Halcon has developed new routes to acetic anhydride (see p. 70) and vinyl acetate (see p. 71) based on methanol, and Monsanto is exploring the feasibility of producing styrene from toluene and methanol using a zeolite-type catalyst. Further down the road are the catalytic cracking of methanol to ethylene and the manufacture of ethylene glycol from methanol. Figure 7 illustrates the scope of the chemistry of syngas via methanol.

Methanol is also used as feedstock for the production of single-cell proteins for animal and human consumption. The ICI single-cell protein process based on methanol uses the bacterium *Methylophilus methylotropus*, and approximately 6000 tons per month of single-cell protein is produced. The organism is grown continuously at 35–42°C, and the reaction products are separated by agglomeration and centrifugation. Purified carbon dioxide is sold as a by-product. The process is depicted in Figure 8.

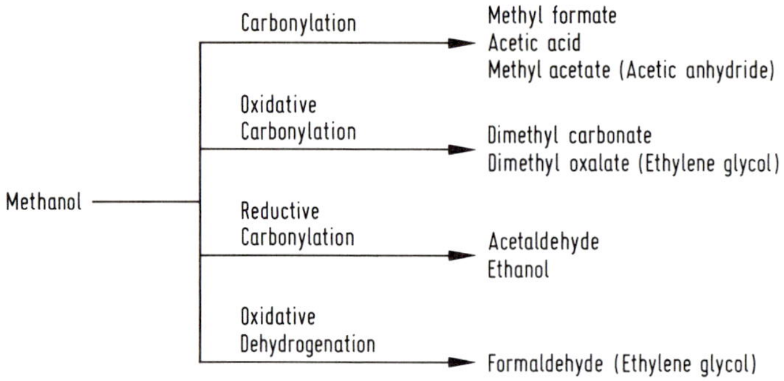

Figure 7. Chemicals derived from methanol and syngas. The reaction products in parentheses are made by subsequent transformations also involving syngas.

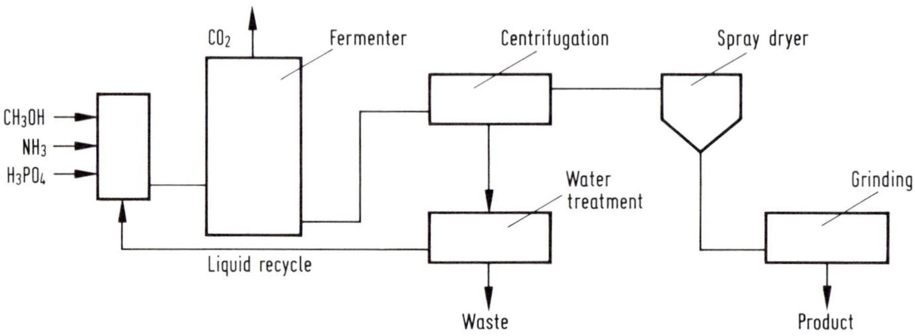

Figure 8. Process for production of ICI single-cell protein from methanol.

Other commercial processes for the production of single-cell protein include algal, yeast, and fungal processes. Usually these processes are limited to regions where low-cost carbon and energy sources are available and conventional animal feedstuff proteins such as soybean meal and fish meal are in short supply. Food grade yeast products are also used as flavoring agents. The possibility of improving these processes by genetic engineering is currently under investigation.

Other important industrial chemicals derived from methanol not described separately include the methylamines, which are manufactured by stepwise methylation of ammonia. In the process used most often, methanol and ammonia are reacted at 350–500°C and 10–200 bar over an aluminum silicate or phosphate catalyst.

$$CH_3OH + NH_3 \longrightarrow CH_3NH_2 + H_2O$$

All three methylamines are usually obtained, and separation is achieved by a combination of pressure and extractive distillation. Dimethylamine is used in the manufacture of *N,N*-dimethylformamide and *N,N*-dimethylacetamide, which are important polar solvents used in the manufacture of polymers. Methylamine is used mainly for the manufacture of dimethyl urea and *N*-methyl pyrrolidone.

3. Formaldehyde HCHO

Consumption (1984):	United States	2,458,000 tons
	Western Europe	3,622,000 tons
	Japan	1,141,000 tons
Production Process:	Dehydrogenation of methanol	
Major Uses:	UF resins (30%), PF resins (25%), polyacetal resins (5%), butanediol (5%), pentaerythritol (5%)	
Price:	\$ 0.15/kg (37% solution)	

Formaldehyde is an unstable colorless gas, which is commercially supplied in 37% aqueous solution, as a solid cyclic trimer (trioxan), and as a solid polymer (paraformaldehyde). Almost all formaldehyde produced is derived from methanol either by oxidative dehydrogenation using silver or copper catalysts, or by oxidation in the presence of Fe_2O_3 and MoO_3.

Silver crystals or nets are used as catalysts in the dehydrogenation reaction.

$$CH_3OH \xrightarrow{\text{Cat.}} HCHO + H_2$$

The above reaction is conducted at 600–720°C, and the hot gases are quickly cooled to 150°C and washed in a counter-flow with water. The aqueous formaldehyde solution (37–42%) is usually stabilized with 1–2% methanol.

In the oxidative dehydrogenation, the generated hydrogen is oxidized to water upon addition of air.

$$CH_3OH + 0.5\ O_2 \xrightarrow{\text{Cat.}} HCHO + H_2O$$

In this process, methanol vapors, together with a large excess of air, are passed over the catalyst bed in a tubular reactor at 350–450°C.

A direct oxidation of methane to formaldehyde has not yet been developed, mainly because of the low rate of reaction below 600°C and the relative instability of formaldehyde above this temperature. However, Celanese obtains formaldehyde as a by-product in the oxidation of propane or propane/butane mixtures.

Formaldehyde is used mainly in the production of formaldehyde resins. Phenol formaldehyde (PF) resins (consumption in the United States, Western Europe, and Japan in 1984: 1,069,000 tons) are used as plywood adhesives, as binder resins for particleboard, as binder resin in glass fiber insulation, and as molding compounds. Urea formaldehyde (UF) resins (United States consumption in 1984: 545,000 tons) are used mainly as particleboard binder resins and for paper and textile treatment. Melamine formaldehyde (MF) resins (United States consumption in 1984: 95,000 tons) are used in dinnerware and as countertop and tabletape coatings.

The reaction of aniline with formaldehyde in the presence of hydrochloric acid affords aniline/formaldehyde resins that are phosgenated to give polymeric isocyanates. MDI (4,4′-diisocyanatodiphenylmethane) is separated from the polymeric isocyanates by distillation and is used in the manufacture of polyurethane elastomers and fibers. The residual polymeric isocyanates are utilized in the manufacture of polyurethane or polyisocyanurate rigid foams (see p. 145).

Chemical uses of formaldehyde include the manufacture of butanediol and pentaerythritol, which are both polyurethane intermediates. Also, isoprene is manufactured from formaldehyde via 4,4-dimethyl-1,3-dioxane.

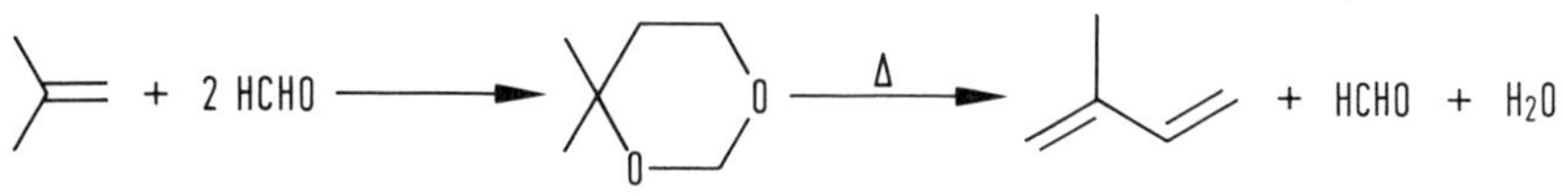

Formaldehyde is also homopolymerized and copolymerized to form polyacetal resins (consumption in the United States, Western Europe, and Japan in 1985: 209,000 tons). The homopolymer is sold by Dupont under the trade name Delrin, while a copolymer with ethylene oxide is sold by Celanese under the trade name Celcon.

4. Phosgene $COCl_2$

Consumption: Not known because phosgene is mainly used captively*
Production Process: Carbon monoxide and chlorine
Major Uses: Diisocyanates (84%), agricultural chemicals (10%), polycarbonates (6%)
Price: $ 0.55 kg

Phosgene is a toxic gas that is readily obtained from chlorine and carbon monoxide in the presence of activated carbon granules as catalyst.

$$CO + Cl_2 \longrightarrow COCl_2$$

In view of its toxicity, phosgene is produced on demand and is usually consumed immediately after formation at the point of manufacture.

The main use of phosgene is in the production of diisocyanates, which are polyurethane building blocks. Phosgenation of oligomeric amines obtained from aniline and formaldehyde affords a mixture of polymeric isocyanates. Subsequent vacuum distillation separates 4,4′-diisocyanatodiphenylmethane (MDI) from the residual polymeric isocyanate (PMDI). Toluene diisocyanate (TDI) is produced by phosgenation of a 80:20 mixture of 2,4-diaminotoluene and 2,6-diaminotoluene. The continuous phosgenation of aromatic amines is conducted at high pressure and elevated temperatures in inert organic solvents such as chlorobenzene or *o*-dichlorobenzene.

Aliphatic diisocyanates such as hexamethylene diisocyanate (HDI), isophorone diisocyanate (IPDI), and hydrogenated MDI are also manufactured by phosgenation of the corresponding diamines. The commercially important isocyanates obtained by phosgenation of amine precursors are listed in Table 14.

Hydrogen chloride is a by-product from the manufacture of diisocyanates, and it is used in the manufacture of vinyl chloride, the monomer for PVC.

Phosgene is also used to manufacture dicarboxylic acid chlorides, which are building blocks for condensation polymers such as polyamides and polyesters. For example, aromatic polyamides (aramides) are made from either isophthaloyl chloride and *m*-phenylenediamine (Nomex) or terephthaloyl chloride and *p*-phenylenediamine (Kevlar).

Polycarbonates are manufactured from phosgene and bisphenol A. This reaction is conducted by interfacial polycondensation, using a mixture of methylene chloride and aqueous sodium hydroxide. Mixtures of phosgene and dicarboxylic acid chlorides are also used in combination with bisphenol A to synthesize polyester/polycarbonate copolymers.

* For the manufacture of diisocyanates, a total of 1,447,000 tons of phosgene was produced worldwide in 1986.

Table 14. Commercial Isocyanates

Name	Structure	Application
TDI	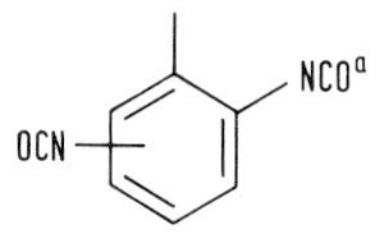	Flexible polyurethane foams
MDI	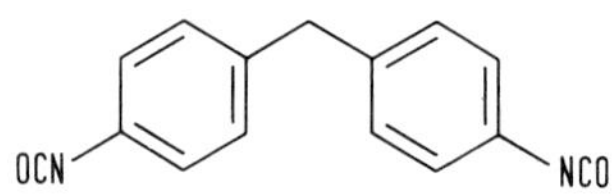	Polyurethane elastomers and fibers
PMDI	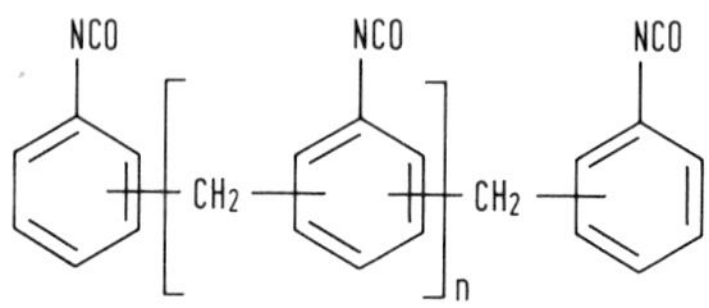	Rigid polyurethane foams
HDI	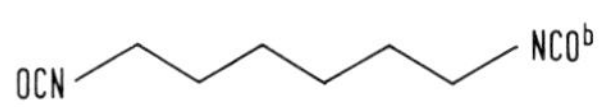	Polyurethane coatings
IPDI	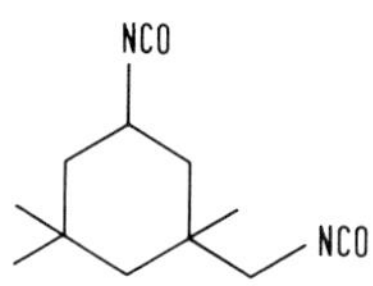	Polyurethane coatings
H_{12}MDI	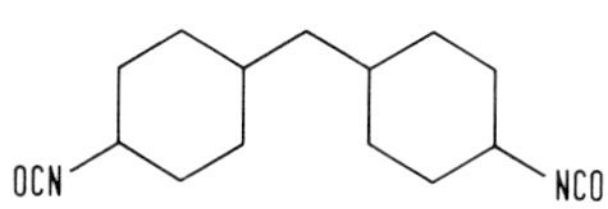	Polyurethane coatings and elastomers

[a] 80:20 mixture of 2,4 and 2,6 isomers.
[b] Used only as a functional derivative (biuret, isocyanurate) in rigid polyurethane coatings.

Attempts to develop nonphosgene processes for the manufacture of diisocyanates and polycarbonates have not been successful. One exception is the new route to methyl isocyanate from *N*-methylformamide pioneered by Dupont. Vapor-phase oxidation of *N*-methylformamide produces the volatile methyl isocyanate, which is consumed immediately in the production of agricultural chemicals.

$$CH_3NHCHO + 0.5\ O_2 \longrightarrow CH_3NCO + H_2O$$

5. Urea H_2NCONH_2

Consumption (1984):	United States	7,148,000 tons
	Western Europe	7,400,000 tons
	Japan	1,401,000 tons
Production Process:	Ammonia and carbon dioxide under pressure	
Major Uses:	Fertilizer (79%), livestock feeds (12%), urea formaldehyde resins (7%), melamine resins (1%)	
Price:	$ 150/ton	

Urea is commercially produced by reacting carbon dioxide with ammonia. This reaction is conducted at 195°C and 25 MPa, using an ammonia/carbon dioxide ratio of 4:1. The reaction proceeds via ammonium carbamate.

$$2\ NH_3 + CO_2 \longrightarrow H_2NCOONH_4$$

$$H_2NCOONH_4 \longrightarrow H_2NCONH_2 + H_2O$$

Urea processes provide an aqueous solution containing 70–80% urea. This solution can be used directly as fertilizer. A combination of crystallization and remelting removes biuret, which is detrimental to the manufacture of urea-derived polymers.

Urea's major use is in solid and liquid fertilizers, and it is also used in animal feed as a protein supplement. Adhesive and plastic uses include the urea formaldehyde resins and melamine formaldehyde resins. Melamine is obtained by heating urea to 350–400°C over an Al_2O_3 catalyst.

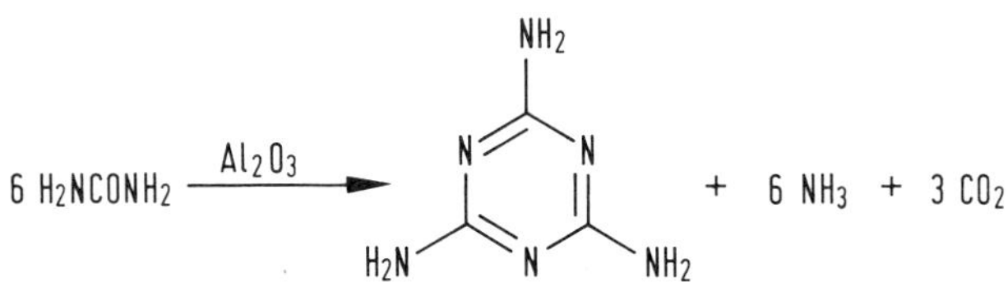

Reaction of melamine with formaldehyde affords *N*-methylol compounds, which are either converted into glassy polymers upon heating or used as the reactants in the formulation of polyurethane polyols.

6. Hydrogen Cyanide HCN

Consumption (1983):	United States	430,000 tons
	Western Europe	280,000 tons
	Japan	72,000 tons
Production Process:	Ammoxidation of methane. By-product in the ammoxidation of propylene to acrylonitrile	
Major Uses:	Adiponitrile (38%), methacrylic acid and esters (33%), cyanuric chloride (9%)	
Price:	$ 1.10/kg	

Hydrogen cyanide is produced mainly by the ammoxidation of methane. In the Andrussow process a catalyst consisting of a 90% platinum/10% rhodium wire gauze is used.

$$CH_4 + NH_3 \xrightarrow{O_2} HCN + 3\ H_2O$$

A variation of this process is practiced by Degussa. In the Degussa process, methane and ammonia are reacted in the absence of oxygen at 1200–1300°C in the presence of platinum, ruthenium, or aluminium catalysts.

$$CH_4 + NH_3 \xrightarrow{Cat.} HCN + 3\ H_2$$

Hydrogen cyanide is also produced as a coproduct in the ammoxidation of propylene to acrylonitrile. In the Sohio process, 12–14 kg of hydrogen cyanide is produced per 100 kg of acrylonitrile. Of the 14 plants in the United States producing hydrogen cyanide, 8 use the direct process and 6 are acrylonitrile units.

Hydrogen cyanide is used in the manufacture of adiponitrile from butadiene (Dupont process). Adiponitrile is an intermediate for nylon-6,6. Also, methacrylates can be synthesized via acetone cyanohydrin (see p. 111). Chlorination of hydrogen cyanide produces cyanogen chloride, which can be trimerized to cyanuric chloride. Cyanuric chloride is a dyestuff intermediate and is also used to manufacture 1,3,5-triazine herbicides.

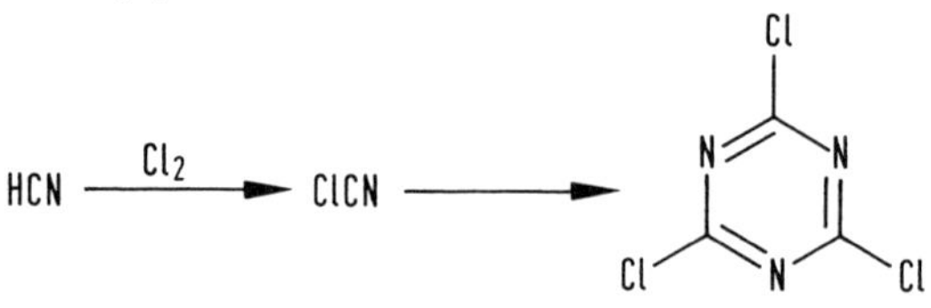

Reaction of hydrogen cyanide with 4,4′-diisocyanatodiphenylmethane affords poly(parabanic acids), which were pilot-produced by Exxon and converted into high-performance films. This reaction proceeds in several steps: formation of linear cyanoformamidines, cyclization to poly(iminoimidazolidinones), and hydrolysis to poly(parabanic acid).

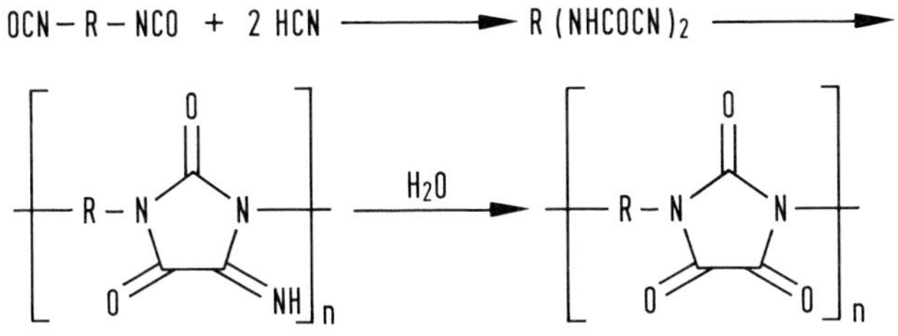

This is the only example where hydrogen cyanide has been used as a co-monomer.

7. Chloromethanes CH_3Cl, CH_2Cl_2, $CHCl_3$, CCl_4

Consumption (1981):		United States	W. Europe	Japan
	CH_3Cl	204,000 tons	193,000 tons	42,000 tons
	CH_2Cl_2	237,000 tons	222,000 tons	31,000 tons
	$CHCl_3$	165,000 tons	87,000 tons	39,000 tons
	CCl_4	286,000 tons	348,000 tons	73,000 tons

Production Process: Chlorination of methane

Major Uses:
- CH_3Cl: Methylchlorosilanes (63%), solvent (8%)
- CH_2Cl_2: Solvent (50%), blowing agent (19%)
- $CHCl_3$: Fluorocarbon 22 (90%)
- CCl_4: Fluorocarbon 12 (58%), fluorocarbon 11 (30%)

Price: CH_3Cl, \$ 0.57/kg; CH_2Cl_2, \$ 0.77/kg; $CHCl_3$, \$ 0.77/kg; CCl_4, \$ 0.53/kg

All four chloromethanes can be obtained by chlorinating methane. However, for chloromethane and tetrachloromethane, direct methods of production are also used. The first industrial gas-phase chlorination of methane was conducted by Hoechst in 1923. Using an equimolar ratio of chlorine and methane, all four chlorinated methanes are obtained:

$$CH_4 + Cl_2 \xrightarrow[-HCl]{440\,°C} \underset{37}{CH_3Cl} + \underset{41}{CH_2Cl_2} + \underset{19}{CHCl_3} + \underset{3}{CCl_4} \quad (37 : 41 : 19 : 3)$$

If a higher degree of chlorination is required, the lower chlorinated products are recycled.

Oxychlorination is a second route to chlorinated methanes. This process has the advantage over the direct chlorination that the waste product HCl is used instead of expensive chlorine.

$$CH_4 + HCl + O_2 \longrightarrow CH_3Cl + CH_2Cl_2 + CHCl_3 + CCl_4 + H_2O$$

The oxychlorination process uses a melt consisting of $CuCl_2$ and KCl as catalyst.

Carbon tetrachloride is also obtained by chlorination of carbon disulfide, propane/propylene mixtures, and high-chlorine-containing residues. Chloromethane is manufactured directly from methanol and HCl.

$$CH_3OH + HCl \longrightarrow CH_3Cl + H_2O$$

This reaction is best conducted in the vapor phase (300–380°C and 3–6 bar) using Al_2O_3 as the catalyst.

Chlorinated methanes are widely used as solvents. The largest use for methyl chloride is in the manufacture of methylchlorosilanes, which are intermediates in the manufacture of silicones. Also, tetramethyllead is manufactured in sizable amounts from methyl chloride. Methylene chloride is used as a propellant for aerosols and as a blowing agent for flexible polyurethane foams. Chloroform and carbon tetrachloride are mainly used in the manufacture of fluorocarbons 11, 12, and 22 (see p. 52), all of which are blowing agents and refrigerants.

8. Fluorocarbons CCl_3F, CCl_2F_2, $CHClF_2$

Consumption (1984):	United States	W. Europe	Japan
CCl_3F, FC-11	84,000 tons	135,000 tons	
CCl_2F_2, FC-12	155,000 tons	180,000 tons	
$CHClF_2$, FC-22	116,000 tons	68,000 tons	150,000 tons[a]
CCl_2FCClF_2, FC-113	150,000 tons	36,000 tons	
Others	70,000 tons	21,000 tons	

Production Processes: Reaction of carbon tetrachloride with hydrogen fluoride (FC-11, 12); reaction of tetrachloroethylene with hydrogen fluoride (FC-113, 114); chloroform and hydrogen fluoride (FC-22)

Major Uses: Air conditioning (30%), blowing agents (18%), refrigeration (16%), solvent (14%), polymer precursor (14%), aerosol (2%)

Prices: FC-11, \$ 1.25/kg; FC-12, \$ 1.50/kg; FC-22, \$ 2.30/kg; FC-113, \$ 2.05/kg; FC-114, \$ 2.22/kg

The industrial process for fluorochloromethanes is the stepwise replacement of chlorine in carbon tetrachloride using hydrogen fluoride. This reaction proceeds in the gas phase at 150°C on a fixed-bed catalyst consisting of the fluorides of aluminum, chromium, or antimony.

$$CCl_4 + HF \longrightarrow CFCl_3 + CF_2Cl_2 + HCl$$

Chloroform is used as the starting material in the manufacture of chlorodifluoromethane (FC-22). Reaction of tetrachloroethylene with hydrogen fluoride produces FC-113 (1,1,2-trichloro-1,2,2-trifluoroethane) and FC-114 (1,2-dichlorotetrafluoroethane).

The fluorochloromethanes are low-boiling, nonflammable, and inert materials that have found uses as propellants for aerosols, as blowing agents for rigid and flexible polyurethane foams, and as refrigerants. The major products used as blowing agents for flexible and rigid polyurethane foams, polystyrene, and polyethylene are FC-11 and FC-12. In the rigid polyurethane insulation foams, the fluorocarbon blowing agents are trapped in the closed cells to provide better insulation than air. Fluorocarbon FC-22 is also used as the starting material for tetrafluoroethylene, and fluorocarbon FC-113 is the precursor of chlorotrifluoroethylene. The alleged depletion of the ozone layer by fluorochloromethanes has led to a 14% drop in their production since 1974. The fluorochloromethanes were replaced as propellants in the United States with less expensive but flammable hydrocarbons (propane, butane, and isobutane). Attempts are under way to replace the fluorochloromethanes with stratospherically safe fluorocarbons containing carbon-hydrogen bonds, which would reduce the lifetime of these molecules in the troposphere.

[a] Total fluorocarbon consumption.

Chapter III. C_2 Chemicals

1. Introduction

Ethylene is the most important building block for industrial organic chemicals as well as for polymers. Approximately half of the ethylene produced is used as a monomer for polyethylene. The other half is used to manufacture three other important monomers for the plastics industry – vinyl chloride, vinyl acetate, and styrene – as well as large-volume chemicals such as ethanol, C_2 chlorinated solvents, acetic acid, ethylene oxide, and ethylene glycol. The most important industrial organic chemicals derived from ethylene are listed in Table 15.

The C_2 chlorinated solvents alone were used to the extent of 1.7 million metric tons in the world in 1980. These solvents – perchloroethylene, 1,1,1-trichloroethane, and trichloroethylene – are used for the degreasing of metals, for dry cleaning, and for the processing of textiles.

Table 15. Industrial Organic Chemicals Derived from Ethylene

Product	United States Consumption, 1983 (1000 tons)
Ethylene dichloride	5,112
Ethylbenzene	3,571
Styrene	3,175
Vinyl chloride	3,159
Ethylene oxide	2,534
Ethylene glycol	2,027
Acetic acid	1,270
Vinyl acetate	891
Acetic anhydride	568
Acetaldehyde	397
Methylchloroform	274
Perchloroethylene	230
Ethanolamines	213

Many important chemical intermediates are obtained by oxidation of ethylene as summarized in Figure 9.

Other important reactions involve addition to reactive double bonds. The alkylation of benzene has been included because addition to a benzene ring produces ethylbenzene, the intermediate for styrene monomer. These reactions are summarized in Figure 10. Recently, it was demonstrated that ethylene reacts with ammonia to give ethylamine. This reaction proceeds at 370°C in the presence of an acidic zeolite catalyst.

Another important C_2 chemical is tetrafluoroethylene, the monomer for polytetrafluoroethylene (Teflon). However, this monomer is produced by C_1 chemistry. Reaction of chloroform with hydrogen fluoride produces chlorodifluoromethane, which is pyrolyzed to difluorocarbene. Difluorocarbene dimerizes under the reaction conditions below to give tetrafluoroethylene.

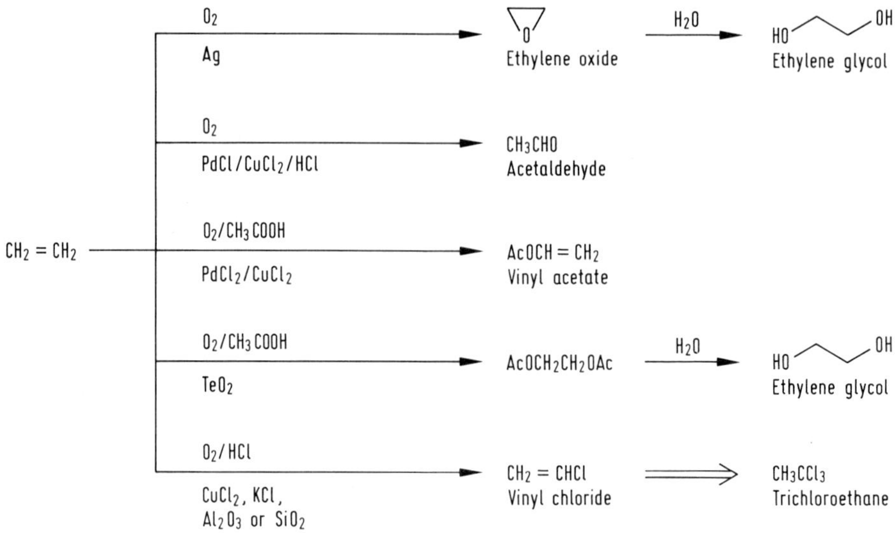

Figure 9. Major products obtained by oxidation of ethylene. Ac = CH_3CO-.

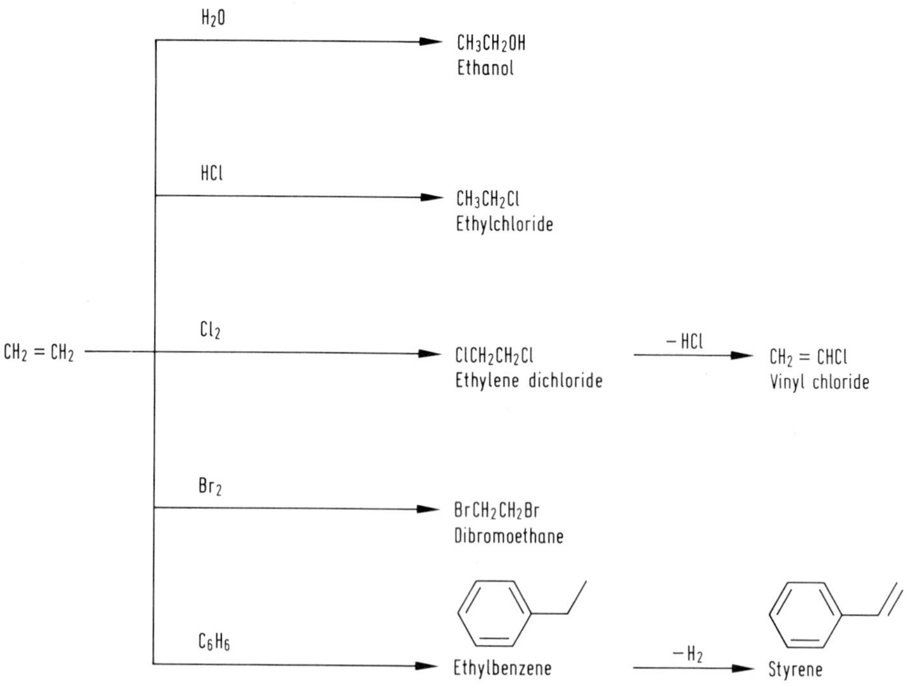

Figure 10. Major products obtained by addition to ethylene.

$$CHCl_3 + 2\,HF \longrightarrow CHClF_2 + 2\,HCl$$

$$CHClF_2 \xrightarrow[600\,°C]{\Delta} [:CF_2] \longrightarrow CF_2 = CF_2 + 2\,HCl$$

The polyethylene industry is in the midst of major changes. New export-oriented Saudi Arabian and Canadian plants have started up, and new plants are on the drawing board in Mexico, Eastern Europe, Southeast Asia, and Africa. For example, petrochemical plants at Yanbu, Saudi Arabia, have capacities of 450,000 tons of ethylene, 200,000 tons of ethylene glycol, and 320,000 tons of linear low- and high-density polyethylene. Linear low-density polyethylene will displace a significant proportion of conventional low-density polyethylene, and the downgauging of film and other products will lower the overall growth rate of polyethylene.

2. Ethylene $CH_2=CH_2$

Consumption (1986):	United States	14,900,000 tons
	Western Europe	12,900,000 tons
	Japan	4,300,000 tons
Production Process:	Thermal (steam) or catalytic cracking of hydrocarbons	
Emerging Technologies:	Steam cracking of crude oil	
Major Uses:	Polyethylene (45%), ethylene oxide (20%), vinyl chloride (15%), styrene (10%)	
Price:	$ 0.42/kg	

In the United States, ethylene was derived mainly from inexpensive natural gas, which is available in large quantity. In contrast, Western Europe and Japan opted for petroleum as feedstock for ethylene. The recent deregulation of natural gas in the United States led to price increases, which prompted a switch to petroleum. Since naphtha is the preferred feedstock for aromatics in the United States, efforts are being made to use the higher boiling gas oil as feedstock for olefins. The unsettled feedstock situation has prompted construction of refineries with built-in flexibility to use natural gas as well as higher boiling liquids. Flexibility requires greater investments. In 1985 a gas cracker cost $ 550 million, a naphtha cracker $ 690 million, and a gas oil plant $ 770 million.

The ethylene feed flexibility in the United States is demonstrated by the fact that in 1980 about 49% of ethylene was produced from gas liquid feedstocks and 42% from heavy liquid feedstocks. In 1985 a total of 41% of ethylene producers used gas liquid feedstocks, and only 20% used heavy liquid feedstocks exclusively. About 39% of the ethylene producers used either feedstock depending on price and availability. The United States trend to heavy liquid cracking favors oil companies with access to these feeds from associated refineries as well as chemical companies that enter into joint ventures with oil companies. A recent example is the merger of Dupont with Conoco. Heavy liquid crackers can be adjusted to change the relative yields of coproducts such as propylene and butadiene. The trend to heavy liquid cracking in the United States can be seen in Table 16. The gas feedstocks yield minimum coproducts. A diagram of a gas cracker is shown in Figure 11.

The slightly lower temperatures used in the naphtha and gas oil crackers allow better control of the relative amounts of valuable coproducts (propylene, butadiene).

The direct cracking of crude oil would free chemical companies from dependence on oil refiners and gas processors for supplies of ethane, propane, naphtha, and gas oil feedstocks.

Union Carbide and Dow have built direct cracking prototype plants. One advantage of this approach is the generation of sizable amounts of acetylene. For example, a 450,000 tons/year plant based on the Dow process would produces 50,000 tons of acetylene.

Table 16. Feedstocks Used in United States Ethylene Plants (1000 tons)

Feedstock	1980	1985	1990[c]
Ethane	6,800	7,500	7,500
Propane	2,700	2,300	1,800
Butane	450	450	360
Refinery gas	180	180	180
Naphtha	4,450	6,000	7,300
AGO[a]	2,900	4,000	6,800
VGO[b]	1,000	1,900	5,000
Total	18,480	22,330	28,940

[a] Atmospheric gas oil, boiling point 175–345°C
[b] Vacuum gas oil, boiling point 345–455°C
[c] Projected

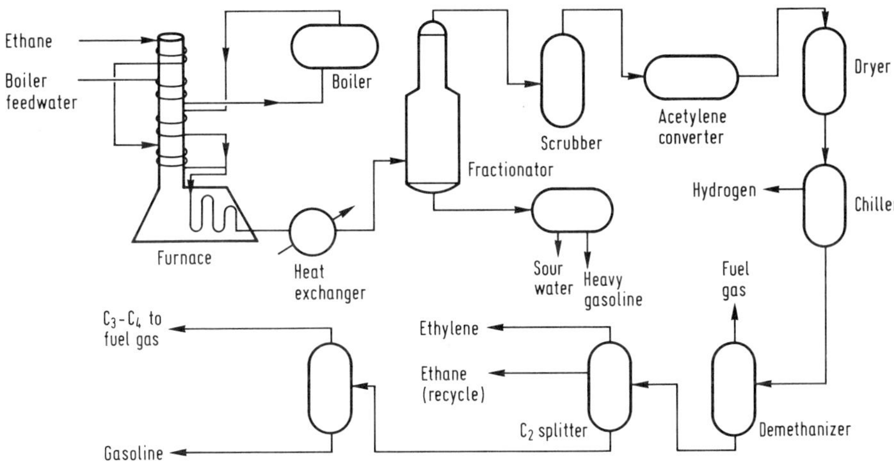

Figure 11. Gas cracker.

Ethylene is the monomer for polyethylene. Three types of polyethylene polymers are produced in the world. Low-density polyethylene (LDPE), which is produced under high pressure and temperature (350°C and 1,000–3,000 bar) using peroxide initiators; linear low-density polyethylene (LLDPE), produced by a low-pressure process; and high-density polyethylene (HDPE), which is produced at 50–160°C and low pressure, using Ziegler-Natta or metal oxide catalysts. The consumption of polyethylene in three major world markets (United States, Western Europe and Japan) in 1986 is shown in Table 17.

The trend in polyethylene production is toward the manufacture of more cost-effective products with improved physical properties. Linear low-density polyethylene (LLDPE) polymers have been introduced. These new LLDPE polymers are linear but have a significant number of branches introduced by using comonomers such as butene-1 or octene-1. The linearity provides strength, while the branching provides toughness. A 1.5-mil (ca. 40 μm) film of LLDPE has the same tear strength as a 5-mil (ca. 130 μm) branched LDPE film. The projected growth rate for HDPE is significantly higher than for low-density grades. This can be attributed largely to the expanding use of high-molecular-weight (HMW) HDPE polymers, which have superior physical properties compared to traditional HDPE products.

Table 17. Consumption of polyethylene 1986 (1000 tons)

	LLDPE	LDPE	HDPE
United States	1,335	2,490	2,765
Western Europe	260	3,920	1,650
Japan	120	1,240	750

Chlorosulfonated polyethylene is an elastomer that can be vulcanized with metal oxides. Dupont sells the product under the trade name Hypalon. Production in the United States is on the order of 20,000 tons/year. Hypalon is used in automotive applications, for wire and cable insulation, and as pond liners. Another chlorinated polyethylene, marketed by Dow under the trade name Tyrin, is manufactured by direct aqueous chlorination of polyethylene; it is a thermoplastic elastomer. Crosslinking by free-radical-initiated vulcanization is also used in some formulations, especially for cable jacketing.

3. Ethylene Oxide 

Consumption (1983):	United States	2,534,000 tons
	Western Europe	2,500,000 tons
	Japan	533,000 tons
Production Process:	Vapor-phase oxidation of ethylene with oxygen or air	
Major Uses:	Ethylene glycol (60%), polyols and surfactants (10%), glycol ethers (10%)	
Price:	\$ 0.80/kg	

Ethylene oxide was first produced by air oxidation of ethylene by UCC in 1937. The most effective catalyst for this reaction is metallic silver in a finely divided layer on a support.

$CH_2{=}CH_2 + 0.5\ O_2 \longrightarrow$ 

In all processes, inhibitors are incorporated that prevent the further oxidation of ethylene oxide to carbon dioxide and water. Often several parts per million of 1,2-dichloroethane is added to the ethylene to generate chlorine, which is absorbed on the silver to deactivate the surface. Today oxygen is used almost exclusively in the oxidation. Typical conditions are 250−300°C and 10−20 bar. Since the plant efficiency in producing ethylene oxide is about 72%, selectivity improvements are needed, especially in the face of rising ethylene prices.

Chem Systems in the United States has developed a process that involves ethylene glycol monoacetate as an intermediate. Pyrolysis could provide ethylene oxide. This process is similar to the Halcon process for the production of ethylene glycol via the mono- and diacetate.

$$CH_2{=}CH_2 + O_2 + AcOH \xrightarrow{TeO_2/I_2} HOCH_2CH_2OAc + H_2O$$

$HOCH_2CH_2OAc \xrightarrow[410°/100\ Torr]{AcONa/SiO_2}$  $+\ CH_3COOH$

65 : 35

Approximately 60% of ethylene oxide is used in the production of ethylene glycol, with much of the conversion done in plants right next to the oxide plants. Ethylene glycol is used mainly as a comonomer for polyesters, but it is also used as antifreeze. In addition, ethylene oxide is used in the manufacture of polyols for the polyurethane market. Homopolymerization of ethylene oxide produces poly(ethylene glycol) (PEG) by a base-catalyzed ring-opening polymerization. Also, block and random copolymers with propylene oxide are being produced. These polyether polyols are nonvolatile and range in viscosity from viscous liquids to waxy solids depending on type and molecular weight.

Other secondary products of ethylene oxide are ethanolamines, glycol ethers, and ethylene carbonate.

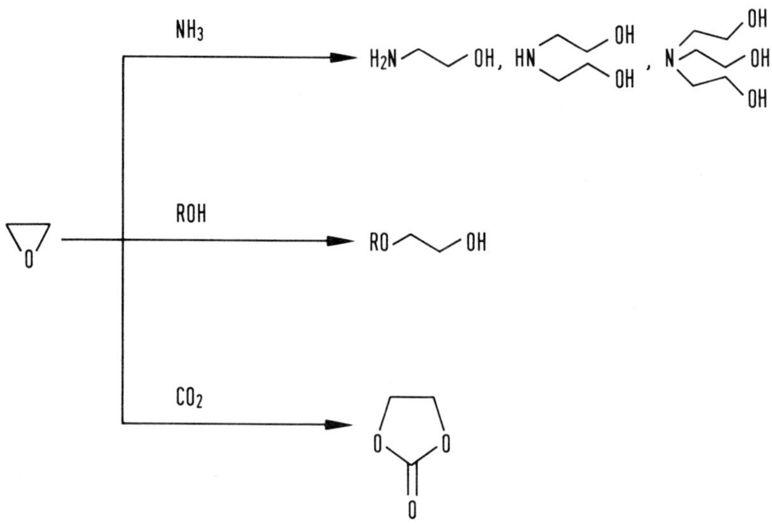

Ethylene oxide reacts exothermally with aqueous ammonia at 60–150°C and 30–150 bar to afford a mixture of the three ethanolamines. Ethanolamines are used in a wide variety of applications ranging from detergents to polyol initiators. They are also used to remove hydrogen sulfide (H_2S) and carbon dioxide (CO_2) from natural gas streams and fuel liquids. Total consumption of ethanolamines in the United States in 1985 was 245,000 tons. Sterically hindered amines are under development for gas treatment. Glycol ethers and ethylene carbonate are used as polar solvents. Further alkylation of the free hydroxyl group in the methyl glycol ether gives the dimethyl ether of glycol (Glyme). The dimethyl ether of diethylene glycol is Diglyme, which is also used as an inert polar solvent.

4. Ethylene Glycol $HOCH_2CH_2OH$

Consumption (1983):	United States	2,181,000 tons
	Western Europe	963,000 tons
	Japan	1,318,000 tons
Production Process:	Hydrolysis of ethylene oxide	
Major Uses:	Polyesters (55%), antifreeze (45%)	
Price:	$ 0.45–0.55/kg	

The most important industrial organic chemical derived from ethylene oxide is ethylene glycol. In a typical process, ethylene oxide is treated with an excess of water and an acidic catalyst at 50–70°C. A higher temperature vapor-phase hydrolysis process (140–230°C and 20–40 bar) is also used.

If the concentration of water is lowered, di-, tri-, and poly(ethylene glycols) are formed. The direct oxidative acetoxylation of ethylene to glycol mono- and diacetate in processes developed by Halcon, Teijin, Kuraray, Celanese, Dupont, and ICI has not yet been commercialized. In the Halcon process, ethylene is converted into a mixture of glycol monoacetate and diacetate using excess acetic acid at 170°C and 28 bar in the presence of tellurium, lithium, and bromide. However, a 360,000-ton/year ethylene glycol plant built by Oxirane never operated because of severe corrosion problems. Direct oxidative hydroxylation would have advantages over acetoxylation, and efforts are being made to optimize this reaction.

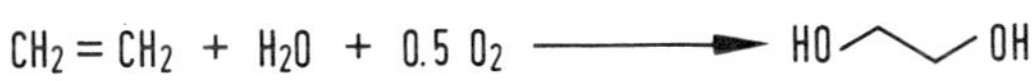

Another process developed by UCC uses synthesis gas as feedstock. The syngas is converted at 125–350°C and 1400–3400 bar in the presence of rhodium carbonyl complexes directly into ethylene glycol. Ube Industries has developed an oxidative carbonylation route to ethylene glycol via oxalates. Hydrocarbonylation of formaldehyde and subsequent reduction of glycolaldehyde is under active investigation.

The major application for ethylene glycol is in antifreeze mixtures in automobile cooler systems. Another important application is its use as a monomer for polyesters. Poly(ethylene terephthalate) is used for fiber, film, and molding applications, and poly(ethylene adipates) are used in the polyurethane industry. For example, consumption of terephthalate polyesters in 1983 in the United States, Western Europe, and Japan were 2,060,000, 889,000, and 669,000 tons, respectively.

Other important secondary products include glyoxal, dioxolane, and dioxane. Glyoxal is only used in aqueous solution because the dialdehyde is not stable in pure form. Dioxolane is a comonomer for polyacetal resins, and dioxane is a widely used polar solvent.

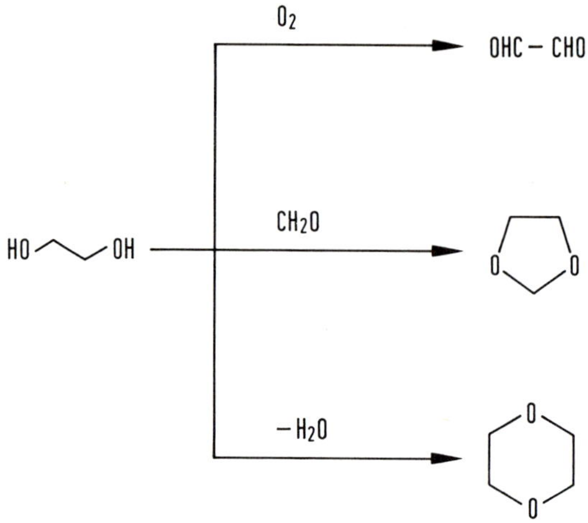
O2
OHC – CHO
CH2O
HO OH
O
O
– H2O
O
O

5. Acetaldehyde CH_3CHO

Consumption (1984):	United States	397,000 tons
	Western Europe	706,000 tons
	Japan	297,000 tons
Production Process:	Oxidation of ethylene (Wacker process); dehydrogenation of ethanol	
Major Uses:	Acetic acid (70%), pyridines (7%), pentaerythritol (5%)	
Price:	$ 0.63/kg	

The Wacker process, the oxidation of ethylene to acetaldehyde involving a homogeneous $CuCl_2/PdCl_2$ catalyst system, was developed in 1957.

$$CH_2{=}CH_2 + 0.5\ O_2 \longrightarrow CH_3CHO$$

In this process $CuCl_2$ reoxidizes palladium metal to the divalent state. The gaseous reagents react with the aqueous catalyst solution in a bubble column reactor (120–130°C and 3 bar).

The second process, oxidation of a mixture of propane and butane (425–460°C and 7–20 bar), was developed by Celanese. However, separation of the by-products formaldehyde, methanol, acetic acid, *n*-propanol, methyl ethyl ketone, and acetone requires a large capital investment. In the United States acetaldehyde is obtained exclusively by oxidizing ethylene (Celanese, Eastman Kodak).

Acetaldehyde is used as a chemical intermediate, and 70% of its production is converted to acetic acid. In recent years the production of acetaldehyde has declined because of a shift to the carbonylation of methanol to produce acetic acid. In Europe, 62% of the acetaldehyde consumption is for the production of acetic acid and 19% is for the manufacture of ethyl acetate.

Several products derived from the aldol condensation of acetaldehyde are commercially important chemicals. The initially formed product, 3-hydroxybutyraldehyde (acetaldol) is easily obtained from acetaldehyde using sodium hydroxide as the catalyst.

Secondary products derived from acetaldol include 1,3-butanediol, crotonaldehyde, *n*-butyraldehyde, and *n*-butanol.

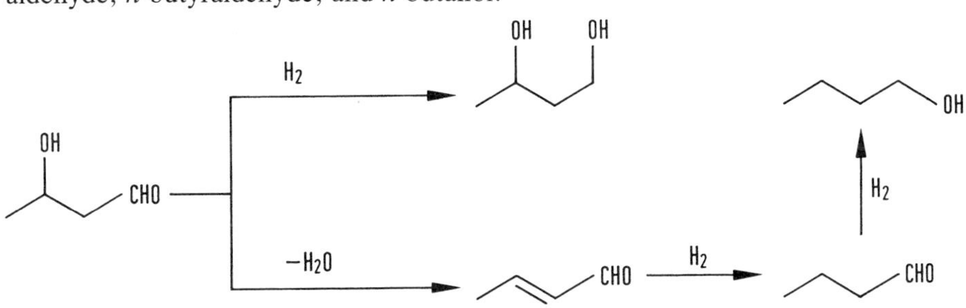

A less costly route for the manufacture of *n*-butyraldehyde and *n*-butanol is the hydroformylation of propylene.

Air oxidation of crotonaldehyde to crotonic acid is readily achieved in the liquid phase (20°C and 3–5 bar). Crotonic acid is used as a comonomer and a leveling agent for alkyd resins.

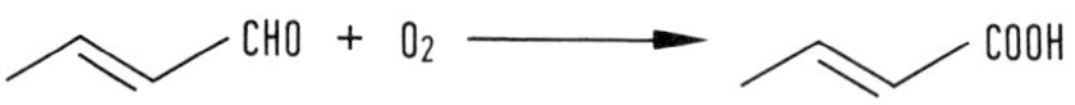

Crotonaldehyde is also used in the manufacture of *trans,trans*-2,4-hexadienoic acid (sorbic acid). Initially, crotonaldehyde reacts with ketene to produce a polyester, which upon hydrolysis yields sorbic acid. The salts of sorbic acid are used as food preservatives.

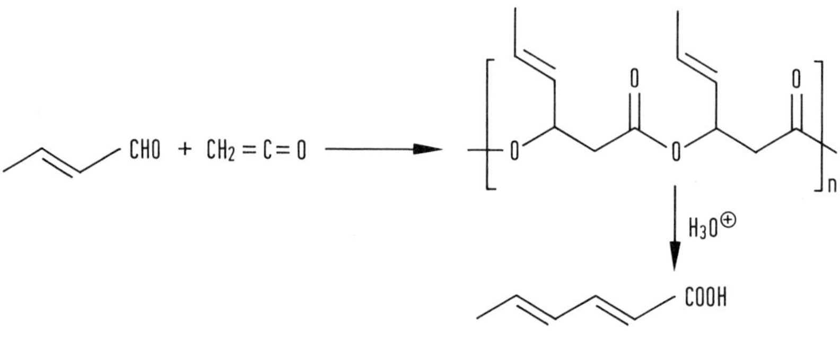

Acetaldehyde is also the basic building block for pyridine and its derivatives. For example, vapor-phase reaction of acetaldehyde and ammonia gives a mixture of 2-methyl- and 4-methylpyridine. If formaldehyde is added, pyridine and 3-methylpyridine are included in the reaction mixture.

2-Methylpyridine is the starting material for 2-vinylpyridine. Reaction of 2-methylpyridine with formaldehyde yields 2-hydroxyethylpyridine, which is subsequently dehydrated to 2-vinylpyridine.

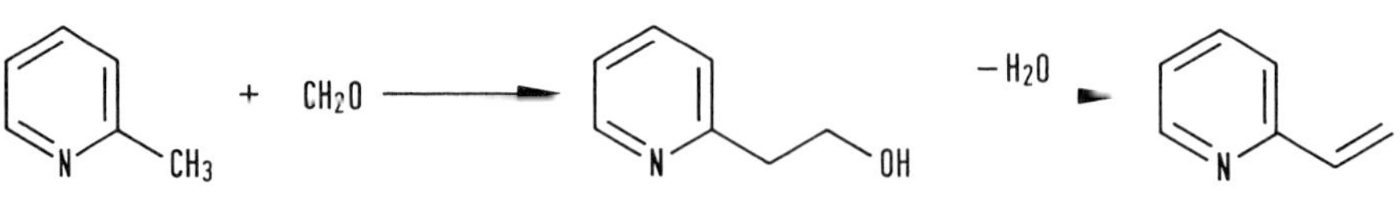

World demand for pyridine and its derivatives in 1982 was on the order of 40,000 tons. Major uses include the herbicide paraquat and terpolymers derived from 2-vinylpyridine, butadiene, and styrene. These terpolymers are used as adhesives for bonding textile fibers to natural and synthetic rubbers in the manufacture of tires and mechanical rubber goods.

6. Acetic Acid CH_3COOH

Consumption (1983):	United States	1,270,000 tons
	Western Europe	1,154,000 tons
	Japan	662,000 tons
Production Process:	Oxidation of acetaldehyde; oxidation of alkanes and alkenes; carbonylation of methanol	
Major Uses:	Vinyl acetate (44%), cellulose acetate (22%), acetic acid esters (10%), solvent (10%)	
Price:	$ 0.55/kg	

Oxidation of acetaldehyde with air yields acetic acid. This reaction proceeds via peracetic acid.

$$CH_3CHO + O_2 \longrightarrow [CH_3CO_3H] \xrightarrow{CH_3CHO} 2\ CH_3COOH$$

In this process peracetic acid can be obtained under mild conditions, and UCC has used peracetic acid to oxidize cyclohexanone to caprolactone, a monomer for polycaprolactone polyols.

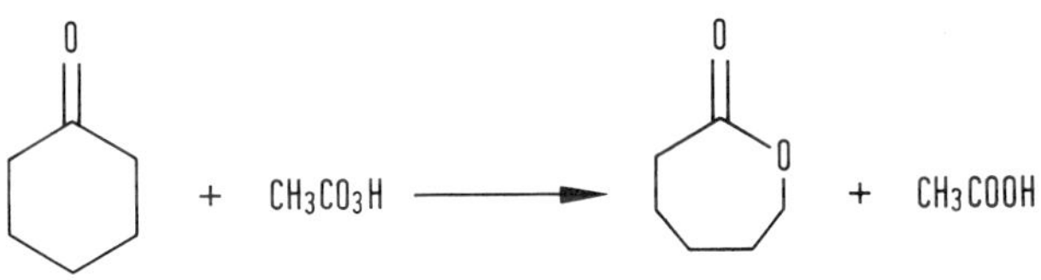

This coproduct process provides acetic acid as by-product.

Oxidation of *n*-butane (Celanese, Hüls, UCC), *n*-butenes (Bayer, Hüls), and light gasoline (B. P., British Distillers) also produces acetic acid. The oxidation of *n*-butane is conducted at 180°C and 15–20 bar, with 10–20% conversion. By-products from the C_4 feedstocks are methyl ethyl ketone and formic acid. The catalyst used is cobalt acetate, but free-radical-initiated oxidation processes are also used. By-products from the C_4–C_8 feedstock (light gasoline) include propionic acid and acetone. Perhaps the most useful future process for acetic acid is the carbonylation of methanol because the feedstock is syngas. In 1985 more than half of the acetic acid produced in the United States was derived from methanol.

$$CH_3OH + CO \longrightarrow CH_3COOH$$

Catalysts in this reaction are CoI_2 (BASF) or, better, RhI_2 (Monsanto). Acetic acid is also produced as a by-product in many reactions. For example, acetic anhydride reactions always produce acetic acid as a by-product.

Acetic acid is a basic industrial organic chemical. It is used in the synthesis of many secondary products, including vinyl acetate, cellulose acetate, butyl acetate, isopropyl acetate, and chloroacetic acid. The largest use of acetic acid is in the production of vinyl acetate. Ethylene is reacted with acetic acid in the vapor phase in the presence of palladium chloride to give vinyl acetate. This monomer is used in the production of poly(vinyl acetate) emulsions and resins, poly(vinyl alcohol), poly-(vinyl butyral), vinyl chloride copolymers, and ethylene vinyl acetate resins. Most of these polymers are used in paints and adhesives.

Shell has developed a syngas-based process for the manufacture of ethyl acetate, a commercial solvent. Hydrocarbonylation of methyl acetate at 200°C and 100 bar produces ethyl acetate and acetic acid for recycle. Ethyl acetate is conventionally produced by reaction of acetic acid with ethanol.

7. Acetic Anhydride $(CH_3CO)_2O$

Consumption (1983):	United States 727,000 tons
Production Process:	Oxidative dehydration of acetaldehyde; dehydration of acetic acid
Emerging Process:	Methyl acetate and carbon monoxide (syngas)
Major Uses:	Cellulose acetate (94%), acetylsalicylic acid (aspirin) (1.4%)
Price:	\$ 0.95/kg

Acetic anhydride is produced by two processes: the modified acetaldehyde oxidation and the dehydration of acetic acid to ketene followed by reaction with acetic acid to the anhydride. The primary acetyl radical obtained by hydrogen abstraction from acetaldehyde is oxidized to the acetyl cation using divalent copper compounds, and subsequent reaction with acetic acid yields the anhydride.

$$CH_3CHO \xrightarrow[-HX]{X^\bullet} CH_3-\dot{C}=O \xrightarrow[-Cu^{I}]{Cu^{II}} CH_3-\overset{+}{C}=O \xrightarrow{CH_3COOH} (CH_3CO)_2O$$

Acetic acid is a by-product in this reaction because peracetic acid obtained in the oxidation of acetaldehyde is used to reoxidize the Cu^{I} species.

The second process, involving ketene as the intermediate, is conducted by treating acetic acid at 700–750°C with a triethyl phosphate catalyst.

$$CH_3COOH \longrightarrow CH_2=C=O + H_2O$$

$$CH_2=C=O + CH_3COOH \longrightarrow (CH_3CO)_2O$$

The higher boiling compounds, acetic anhydride, acetic acid, and water, are separated from the gaseous ketene, and the purified ketene is fed directly into acetic acid at 45–55°C.

Ketene can also be obtained from the pyrolysis of acetone; however, acetone is more expensive than acetic acid, and therefore this route is of minor importance.

The major applications of acetic anhydride include the manufacture of acetyl cellulose. Cellulose acetate and triacetate fibers are made from wood pulp and acetic anhydride. Uses of acetate fibers include women's apparel and cigarette filters. Other uses of acetic anhydride include acetylsalicylic acid (aspirin) and acetanilides.

Ketene is also used to produce diketene. The dimerization occurs at 35–40°C.

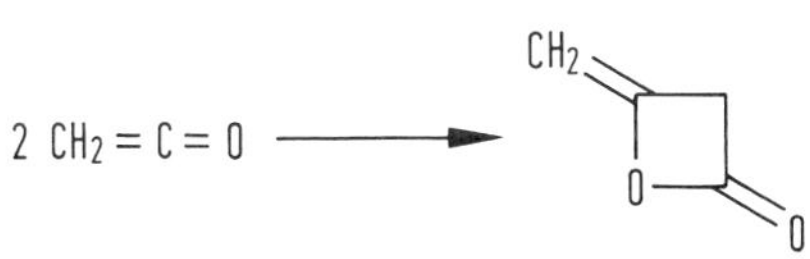

Diketene is used to synthesize acetoacetic acid derivatives. Acetic acid anilides are used as dyestuff and pharmaceutical intermediates.

In 1980 Halcon announced a new process for acetic anhydride. In this process 2 moles of methyl acetate is reacted with carbon monoxide to produce 2 moles of acetic anhydride. One-half the anhydride is recycled and reacted with methanol to give methyl acetate and water. Thus the overall reaction is methanol plus carbon monoxide to yield acetic anhydride and water.

$$2\ CH_3COOCH_3 + 2\ CO \longrightarrow 2\ (CH_3CO)_2O$$

$$(CH_3CO)_2O + 2\ CH_3OH \longrightarrow 2\ CH_3COOCH_3 + H_2O$$

Since the methanol is also derived from synthesis gas, acetic anhydride is completely produced from syngas.

Halcon and Tennessee Eastman have combined technologies, and a 227,000-ton/year plant was recently started up to produce acetic anhydride partly from coal-derived syngas. In the Eastman process, acetic acid obtained from cellulose ester manufacturing is esterified with methanol, and the methyl acetate is reacted with carbon monoxide to give acetic anhydride.

8. Vinyl Acetate $CH_2=CH-OCOCH_3$

Consumption (1984):	United States	653,000 tons
	Western Europe	572,000 tons
	Japan	456,000 tons
Production Process:	Vapor-phase reaction of ethylene, acetic acid, and oxygen	
Major Uses:	Polymers (55%), poly(vinyl alcohol) (20%), copolymers (15%)	
Price:	$ 0.87/kg	

In 1960, all vinyl acetate was produced from acetylene and acetic acid. Acetylene was obtained from calcium carbide, and acetic acid was manufactured by oxidation of ethanol. Wacker simplified the production of acetic acid by oxidizing ethylene to acetaldehyde. More important ICI developed new technology to oxidize ethylene in the presence of acetic acid directly to vinyl acetate (acetoxylation).

$$CH_2=CH_2 + CH_3COOH \xrightarrow[\text{Cat.}]{O_2} CH_2=CH-OCOCH_3$$

The catalyst used was a palladium/copper redox system similar to the one used in the hydroxylation of ethylene (Wacker process). By the end of the 1960s the ethylene route was dominant because of the availability of inexpensive ethylene.

In the 1970s the low-pressure synthesis of ethanol pioneered by ICI and the carbonylation of methanol to acetic acid emerged (Monsanto process). Price increases in oil and natural gas in the 1980s made ethylene too expensive a feedstock for acetic acid, and methanol became the feedstock of choice. However, ethylene is still used for the vinyl portion of the molecule.

The presently used vapor-phase process for vinyl acetate (175–200°C and 5–10 bar) was developed by Bayer and USI. The vapor-phase process minimizes corrosion, and vinyl acetate is formed in high selectivity. The palladium metal catalyst is synthesized by reducing a palladium salt impregnated on a solid support.

Recently Halcon announced the reaction of methyl acetate with carbon monoxide and hydrogen to give ethylidene diacetate and acetic acid. Pyrolysis of ethylidene diacetate produces vinyl acetate and acetic acid.

$$CH_3COOCH_3 + CO + H_2 \longrightarrow CH_3CH(OCOCH_3)_2 + CH_3COOH$$

$$CH_3CH(OCOCH_3)_2 \xrightarrow{\Delta} CH_2=CH-OCOCH_3 + CH_3COOH$$

Since all the acetic acid is recycled to the esterification step, the net reaction amounts to methanol plus carbon monoxide plus hydrogen to give vinyl acetate and water.

Vinyl acetate is an important monomer for the manufacture of a wide variety of polymers. Major uses of these polymeric products are in adhesives and paints. Hydrolysis of poly(vinyl acetat) gives poly(vinyl alcohol). Poly(vinyl alcohol) is used in textile sizes, particularly for polyester-cotton blends. Poly(vinyl butyral), the acetal derived from poly(vinyl alcohol) and butyraldehyde, is used mainly in the laminating of glass (safety glass).

The consumption of poly(vinyl acetate) (PVA) and poly(vinyl alcohol) (the hydrolysis product of PVA) in the three major world markets in 1984 is shown in Table 18. The higher usage of PVAL in Japan is due to fiber manufacture. Vinal or vinylon fibers, produced in Japan, are formaldehyde-modified poly(vinyl alcohol) fibers used in industrial covering applications.

Table 18. Consumption of Poly(vinyl acetate) (PVA) and Poly(vinyl alcohol) (PVAL), 1984

	Consumption (tons)	
	PVA	PVAL
United States	430,000	68,000
Western Europe	380,000	57,000
Japan	71,000	118,000

9. Vinyl Chloride $CH_2=CHCl$

Consumption (1983):	United States	3,159,000 tons
	Western Europe	2,500,000 tons
	Japan	1,420,000 tons
Production Process:	Dehydrochlorination of 1,2-dichloroethane; acetylene and HCl	
Major Uses:	Vinyl polymers	
Price:	\$ 0.46 kg	

Vinyl chloride monomer is used almost exclusively in the manufacture of homo- and copolymers. Poly(vinyl chloride) has the advantage over other thermoplastic polyolefins of built-in fire retardancy because it is 57% chlorine.

Vinyl chloride is manufactured by the dehydrochlorination of 1,2-dichloroethane obtained by the addition of chlorine to ethylene or by oxychlorination of ethylene with gaseous HCl and air (220–240°C and 2–4 bar).

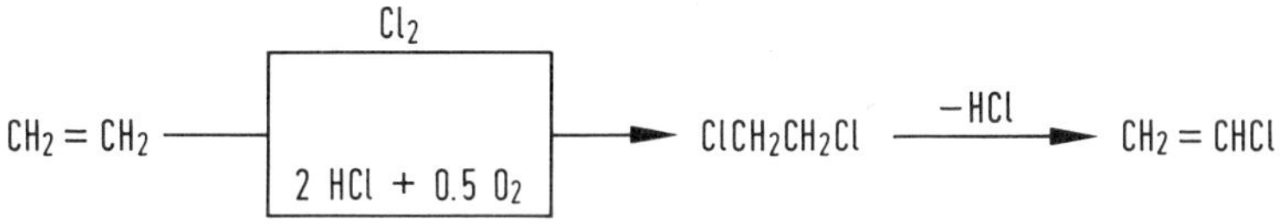

The dehydrochlorination of 1,2-dichloroethane is conducted in the gas phase (500–600°C and 25–35 bar). The conversion is about 50–60% with a selectivity of 98%. The two processes to 1,2-dichloroethane are integrated to use the HCl generated in the thermolysis. Previously vinyl chloride was manufactured by a balanced ethylene/acetylene process; hydrogen chloride obtained in the thermolysis of 1,2-dichloroethane was added to acetylene.

$$HC\equiv CH + HCl \xrightarrow{HgCl_2} CH_2=CHCl$$

Vinyl chloride monomer is usually polymerized by a free radical process to give poly(vinyl chloride) (PVC). The consumption of PVC in 1982 in the United States was 2,574,000 tons; in Western Europe 3,703,000 tons; and in Japan 1,322,000 tons. Vinyl chloride has also been used to manufacture vinylidene chloride and other chlorinated C_2 products.

Vinylidene chloride is manufactured by chlorinating vinyl chloride to 1,2,2-trichloroethane followed by dehydrochlorination over $Ca(OH)_2$ or NaOH at 100°C.

$$CH_2=CHCl \xrightarrow{Cl_2} ClCH_2CHCl_2 \xrightarrow{-HCl} CH_2=CCl_2$$

Vinylidene chloride monomer is homo- or copolymerized to produce polymers used in films and coatings. The consumption of poly(vinylidene chloride) in the United States in 1984 was 86,000 tons.

Vinyl fluoride and vinylidene fluoride are also used as monomers for homo- and copolymers. Vinyl fluoride is manufactured by adding hydrogen fluoride to acetylene.

$$HC \equiv CH + HF \longrightarrow CH_2=CHF$$

This reaction is best conducted in the presence of heterogeneous catalysts, such as mercury compounds or aluminum, tin, or zinc fluorides at 40–400°C. Another route to vinyl fluoride involves the addition of HF to vinyl chloride followed by dehyrochlorination.

$$CH_2=CHCl + HF \longrightarrow CH_3CHClF \xrightarrow{-HCl} CH_2=CHF$$

Poly(vinyl fluoride) is a relatively expensive polymer that has not found widespread applications. Vinylidene fluoride can be obtained by elimination of HCl from 1,1-difluoro-1-chloroethane.

$$CH_3CClF_2 \xrightarrow{-HCl} CH_2=CF_2$$

Trichloro- and tetrachloroethylene are important organic solvents. Chlorination of 1,2-dichloroethane in the gas phase proceeds with simultaneous dehydrochlorination if the reaction is conducted at 350–450°C on a fixed-bed catalyst containing $CuCl_2$, $FeCl_3$, or $AlCl_3$.

$$ClCH_2CH_2Cl \xrightarrow[-HCl]{Cl_2} Cl_2C=CHCl + Cl_2C=CCl_2$$

There is also a modified process that utilizes the generated HCl in an oxychlorination/dehydrochlorination sequence. Trichloroethylene is used as dry cleaning solvent.

10. Ethanol C_2H_5OH

Consumption (1983):	United States	501,000 tons*
	Western Europe	977,000 tons
	Japan	84,000 tons
Production Process:	Hydration of ethylene (70%), fermentation of carbohydrates (27%)	
Emerging Process:	Methanol homologation	
Major Uses:	Solvent (55%)	
Price:	$ 0.33/liter	

Ethanol is manufactured from ethylene by direct catalytic hydration. Catalysts used are H_3PO_4/SiO_2 combinations (300°C and 70 bar).

$$CH_2=CH_2 + H_2O \longrightarrow C_2H_5OH$$

Commercially important esters of ethanol are ethyl acetate and ethyl chloride.

Alcohols can be used in internal combustion engines either as fuels or as fuel extenders blended with gasoline. Ethanol is an excellent fuel, but it has only about two-thirds the energy content of gasoline. While ethanol is effective in the Otto-cycle engine, it is not as useful for diesel engine operation; vegetable oils are more useful as diesel fuel additives. Brazil's Proalcohol program was established in 1975. Between 1975 and 1978 ethanol production in Brazil rose from 600 million liters to 1.5 billion liters, and in 1980 approximately 2.5 billion liters of ethanol was produced. A total of 250,000 cars in Brazil ran on ethanol in 1980. The use of wood, especially wood residues and wastes, has an excellent potential for producing large quantities of ethanol or methanol from tree farms and managed forests. The use of cassava roots is also being considered in Brazil to supplement sugarcane as ethanol feedstock, especially between sugarcane harvests. Figure 12 illustrates a typical distillery operation for the production of ethanol from sugarcane.

Sugarcane stalks are crushed in a series of roller mills to extract the juice, which is fermented in large vats to produce a broth containing about 7% ethanol. Distillation through a column produces 95–99% pure ethanol. Nearly all the energy necessary for the fermentation and distillation processes can be generated by burning the dried cane stalk residue (bagasse).

The direct production of ethanol from synthesis gas is also feasible, but much process refinement is needed. However, homologation of methanol with synthesis gas is another viable route. Shell has developed a cobalt-based catalyst that achieves yields of over 85% at 200°C and 150 bar.

$$CH_3OH + 2\ H_2 + CO \longrightarrow CH_3CH_2OH + H_2O$$

Another catalytic process, using iron pentacarbonyl and a tertiary amine cocatalyst, yields gaseous carbon dioxide as a by-product.

* Synthetic.

$$CH_3OH + H_2 + 2\,CO \longrightarrow CH_3CH_2OH + CO_2$$

Commercialization of these processes depends on the availability of inexpensive methanol.

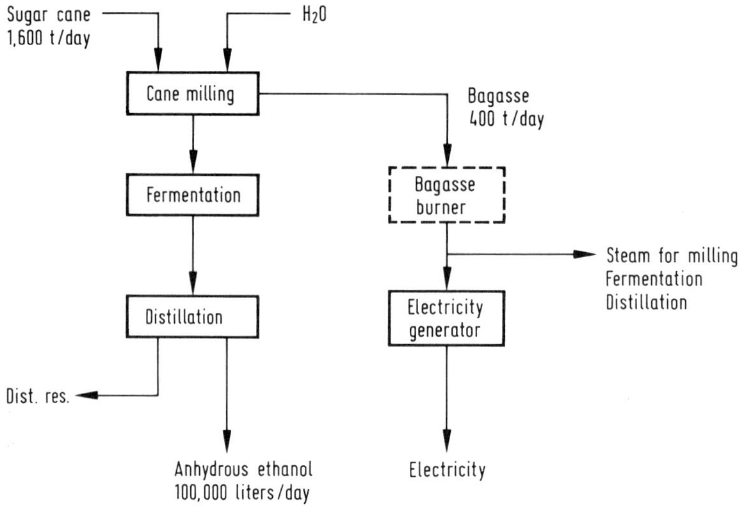

Figure 12. Typical distillery operation to produce ethanol from sugarcane.

11. Ethyl Chloride C_2H_5Cl

Consumption (1983):	United States	128,000 tons
Production Process:	Chlorination of ethane; hydrochlorination of ethylene	
Major Uses:	Tetramethyllead	
Price:	\$ 0.53/kg	

The most economical process for the manufacture of ethyl chloride is the chlorination of ethane and utilization of the hydrogen chloride in the hydrochlorination of ethylene.

$$CH_3CH_3 \xrightarrow[-HCl]{Cl_2} CH_3CH_2Cl$$

$$CH_2=CH_2 \xrightarrow{HCl} CH_3CH_2Cl$$

Hydrochlorination of ethylene is also attractive, depending upon the price of ethylene. This reaction is conducted in the liquid phase at 30–90°C and 3–5 bar with $AlCl_3$ or $FeCl_3$ catalysts or in the vapor phase at 130–250°C with the same catalysts.

Ethyl chloride was formerly used primarily in the manufacture of tetraethyllead, but production of the antiknock agent is decreasing dramatically in the United States because of governmental restrictions.

12. Ethylene Dichloride $ClCH_2CH_2Cl$

Consumption (1983):	United States	5,112,000 tons
	Western Europe	5,789,000 tons
	Japan	2,134,000 tons
Production Process:	Ethylene and chlorine	
Major Uses:	Vinyl chloride (87%), trichloroethane (4%), vinylidene chloride (2%), perchloroethylene (1.5%), trichloroethylene (1.5%)	
Price:	$ 0.37/kg	

Ethylene dichloride is produced from ethylene and chlorine either by chlorination or by oxychlorination. The oxychlorination process is attractive for those with a supply of by-product HCl. For the addition of chlorine to ethylene, catalytic liquid- or vapor-phase reactions of ethylene with chlorine at about 60°C are used.

$$CH_2=CH_2 + Cl_2 \longrightarrow ClCH_2CH_2Cl$$

The chlorides of iron, aluminum, copper, and antimony are used as catalysts. Most plants use ferric chloride in a liquid-phase process. Oxychlorination of ethylene is conducted at 220°C using a fluidized bed or fixed-bed reactor.

$$CH_2=CH_2 + O_2 + 4\ HCl \longrightarrow 2\ ClCH_2CH_2Cl + 2\ H_2O$$

The catalyst is a mixture of copper chloride and other chlorides on a solid support.

13. Chlorinated C_2 Solvents $Cl_2C=CCl_2$, $Cl_2C=CHCl$, Cl_3CCH_3

Consumption (1983):	United States	
	Perchloroethylene ($Cl_2C=CCl_2$)	230,000 tons
	Trichloroethylene ($Cl_2C=CHCl$)	123,000 tons
	Trichloroethane (Cl_3CCH_3)	274,000 tons
Production Process:	Chlorination of ethylene, ethane, and propane	
Major Uses:	Solvents	
Price:	Perchloroethylene, \$ 0.63/kg; trichloroethylene, \$ 0.44/kg; trichloroethane, \$ 0.55/kg	

The basic feedstocks for the chlorinated C_2 solvents perchloroethylene, trichloroethylene, and trichloroethane are ethylene, ethane, and propane. The key intermediate is ethylene dichloride, which is also the starting material for vinyl chloride. Chlorination of ethylene dichloride yields a mixture of perchloroethylene and trichloroethylene in any proportions depending upon the ratio of chlorine to ethylene dichloride.

The PPG trichloroethane process involves vinylidene chloride as an intermediate utilizing the generated HCl in the last step.

$$ClCH_2CH_2Cl + Cl_2 \longrightarrow ClCH_2CHCl_2 + HCl$$
$$ClCH_2CHCl_2 + NaOH \longrightarrow CH_2=CCl_2 + NaCl$$
$$CH_2=CCl_2 + HCl \longrightarrow CH_3CCl_3$$

The Dow process uses vinyl chloride as a key intermediate.

$$ClCH_2CH_2Cl \longrightarrow CH_2=CHCl + HCl$$
$$CH_2=CHCl + HCl \longrightarrow CH_3CHCl_2$$
$$CH_3CHCl_2 + Cl_2 \longrightarrow CH_3CCl_3 + HCl$$

Chlorination of ethane affords trichloroethane (Vulkan process), whereas chlorination of propane produces mixtures of perchloroethylene and carbon tetrachloride.

About 80% of trichloroethylene and 1,1,1-trichloroethane in the United States is used in metal cleaning. Until the early 1970s trichloroethylene was the most widely used vapor degreaser. Because of its toxicity and smog-producing characteristics, its use has been restricted, shifting this market to 1,1,1-trichloroethane. The largest application for perchloroethylene is its use as a dry cleaning solvent. Significant quantities of perchloroethylene are also consumed in the manufacture of fluorocarbons.

14. Acetylene $HC \equiv CH$

Consumption (1983):	United States	116,000 tons
	Western Europe	420,000 tons
	Japan	115,000 tons
Production Process:	By-product in the cracking of hydrocarbons; partial oxidation of methane; hydrolysis of calcium carbide	
Major Uses:	Vinyl chloride (40%), acetylenic chemicals* (28%), vinyl acetate (19%), acrylic acid and esters (5%)	
Price:	$ 0.90/kg	

The important role of acetylene as feedstock for numerous industrial chemicals is based on its remarkable reactivity. However, in recent years the less costly ethylene has replaced acetylene, especially in the production of the high-volume monomers vinyl chloride and vinyl acetate. In 1965, total acetylene consumption in the United States was 524,000 tons. This decreased to 116,000 tons in 1983 (significant amounts of acetylene are consumed in industrial uses as a fuel in cutting and welding).

Acetylene is a coproduct in the manufacture of ethylene, and 87% of the acetylene manufactured in the United States is derived from hydrocarbon feedstock. In contrast, calcium carbide is still being used extensively in Europe and Japan to manufacture acetylene. The most commonly used hydrocarbon feedstocks are natural gas liquids, naphtha, and gas oil, and the yield of acetylene by-product depends on the composition of feed and the severity of cracking. The yields obtained from naphtha or gas oil feedstocks are on the order of 1–2.5%. The acetylene is removed from ethylene by absorption in DMF.

Partial oxidation of methane produces up to 24% acetylene, with carbon monoxide and hydrogen as by-products (BASF process). When the ratio of oxygen to methane is adjusted to 1:2 to achieve incomplete combustion, the by-product is syngas.

$$2\,CH_4 \longrightarrow HC \equiv CH + 3\,H_2$$

A similar process (Wulff process) operates on the pyrolysis of hydrocarbon feedstocks at 1200°C. The energy required for this reaction can also be supplied in the form of an electric arc (Hüls process).

The classical process for the manufacture of acetylene is the reaction of calcium carbide with water. Calcium carbide is obtained from coke and limestone at high temperatures (2200–2300°C).

$$CaC_2 + 2\,H_2O \longrightarrow HC \equiv CH + Ca(OH)_2$$

Historically, vinyl chloride was produced from acetylene and hydrogen chloride in the presence of mercuric chloride.

* Mainly 1,4-butanediol, vinyl ethers, and propargyl alcohol.

$$HC\equiv CH + HCl \longrightarrow CH_2{=}CHCl$$

Today only one plant in the United States uses this technology; all others have switched to ethylene as feedstock (see p. 73). Prior to 1965 acetylene was also the main raw material for vinyl acetate. Today only 6% of the vinyl acetate produced in the United States is based on acetylene. Also, acrylic acid and its esters are now produced from propylene rather than acetylene (see p. 94).

The acetylenic chemicals with growth potential include 1,4-butanediol (see p. 114) and acetylene black, which is produced by burning acetylene. Acetylene black is used in the manufacture of dry cell batteries and as filler in the compounding of plastics and rubbers. Acetylene is also used to produce vinyl fluoride.

$$HC\equiv CH + HF \longrightarrow CH_2{=}CHF$$

Vinyl fluoride is the monomer of poly(vinyl fluoride), which is sold by DuPont as Tedlar film.

Chapter IV. C_3 Chemicals

1. Introduction

Propylene became a popular feedstock for industrial organic polymers after the discovery of its stereoregular polymerization in 1953. Today about 25% (1,882,000 tons) of the propylene in the United States is used as a monomer for polypropylene. Another important monomer, acrylonitrile, accounts for 15% (1,144,000 tons) of the propylene consumption. Propylene is obtained as an ethylene coproduct from natural gas liquids or as a by-product from petroleum refineries and catalytic crackers. Large-volume industrial organic chemicals derived from propylene are listed in Table 19, with data on United States consumption in 1983.

Table 19. Commercially Important Products Derived from Propylene

	United States Consumption, 1983 (1000 tons)
Cumene	1 504
Acrylonitrile	975
Acetone	851
Propylene oxide	822
Isopropanol	550
Acrylic acid and esters	373
Butanol	364
Propylene	218
Glycerol	137

The alkylation of benzene with propylene to cumene is used almost exclusively to produce phenol. In this air-oxidation process, acetone is obtained as a coproduct. Acetone can also be manufactured by oxidation of isopropanol, another important chemical derived from propylene. Major chemical products obtained in the oxidation of propylene are shown in Figure 13.

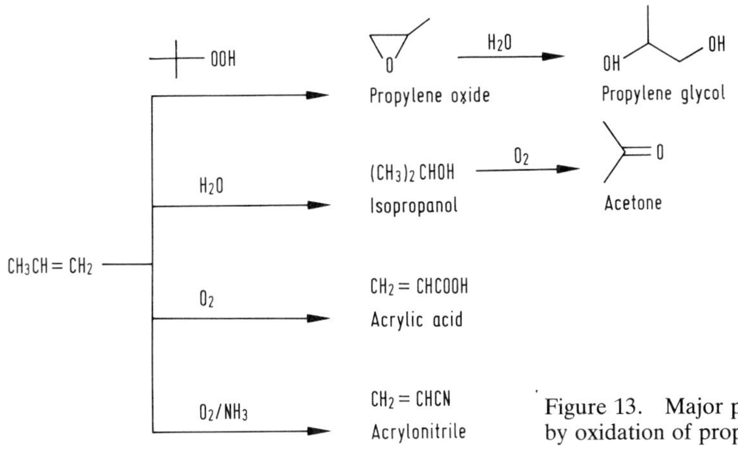

Figure 13. Major products obtained by oxidation of propylene.

Another important reaction of propylene is the oxo process giving rise to the formation of butyraldehyde and the butyraldehyde derivatives 2-ethylhexanol and *n*-butanol.

Oligomerization of propylene accounts for about 8% of propylene consumption. Dodecene (propylene tetramer) and nonene (propylene trimer) are obtained in this manner. The reaction is conducted in the presence of a solid phosphoric acid catalyst. The higher-boiling propylene oligomers (polygas chemicals) are used to increase gasoline yield from petrochemical feedstocks. Nonene is also used in the manufacture of isodecyl alcohol and nonylphenol. Dodecene is used in the manufacture of dodecylbenzene, a common detergent builder. Tridecyl alcohol and dodecylphenol are also produced from dodecene.

2. Propylene $CH_3CH = CH_2$

Consumption (1984):	United States	7,072,000 tons
	Western Europe	5,869,000 tons
	Japan	2,981,000 tons
Production Process:	Thermal (steam) or catalytic cracking of hydrocarbons	
Major Uses:	Polypropylene (25%), acrylonitrile (15%), isopropanol (10%), propylene oxide (10%)	
Price:	$ 0.40/kg	

The shift in feedstocks away from natural gas may in fact produce more propylene in the years to come. When feeds include propane and butane, it is worthwhile to recover propylene and butadiene. The propylene yield can be increased by lowering cracking temperatures (lower conversion). In a typical propane gas cracker, about 375 kg of propylene is produced for 1000 kg ethylene. The propylene yield can be improved to 593 kg per 1000 kg of ethylene, lowering the conversion to 70%. Naphtha and gas oil crackers are producing 420 and 557 kg, respectively, of propylene per 1000 kg of ethylene. Further yield increases can be achieved here also by lowering the conversion. Crude oil crackers produce significantly less propylene. Typical products formed in liquid crackers are listed in Table 20.

Table 20. Typical Products (in weight units) Obtained in Liquid Crackers

	Naphta	AGO[a]
Ethylene	1000	1000
Propylene	420	557
C_4 Fraction	248	412
Butadiene	141	187
Gasoline	741	794
Aromatics	500	409
Fuel oil	195	729
Residues gas	521	457
Total feedstock	3126	3951

[a] AGO = Atmospheric gas oil.

Propylene is used as a monomer to produce polypropylene, which is used in plastics and fiber applications. Polypropylene has a reasonable cost and a good combination of physical and chemical properties. The consumption of polypropylene in the major world markets in 1984 is summarized in Table 21.

Table 21. Manufacture of Polypropylene and Copolymers, 1984 (1,000 tons)

United States	2245
Western Europe	1250
Japan	900

Ethylene propylene (EP) and ethylene propylene diene (EPDM) elastomers are also produced in part from propylene.

Hydration of propylene via the sulfuric acid ester or by direct gas-phase hydration (180–270°C and 25–250 bar) produces isopropanol, which is used as a solvent or as a starting material for acetone.

$$CH_3CH{=}CH_2 + H_2O \longrightarrow (CH_3)_2CHOH$$

3. Propylene Oxide

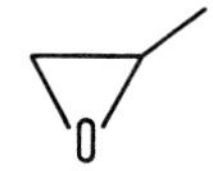

Consumption (1985):	United States	881,000 tons
	Western Europe	826,000 tons
	Japan	190,000 tons
Production Process:	Chlorohydration of propylene; peroxidation of propylene (coproduct process)	
Major Uses:	Polyols (60%), propylene glycols and ethers (29%)	
Price:	\$ 0.80/kg	

The classical chlorohydrin process is still used to manufacture propylene oxide. Dow is the only company in the world that still uses this route. Arco, the other propylene oxide producer, uses a peroxidation process.

Propylene chlorohydrin is produced from propylene, water, and chlorine. Dehydrochlorination to propylene oxide is conducted with calcium hydroxide. The use of sodium hydroxide instead of calcium hydroxide would allow recycling of the aqueous sodium hydroxide to a chlorine plant.

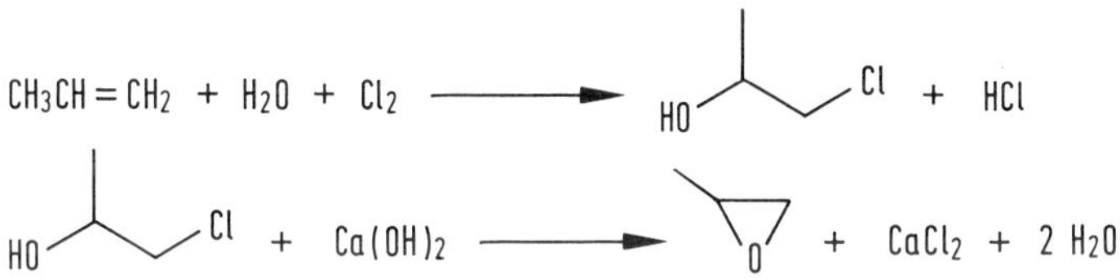

The second process for propylene oxide manufacture, introduced by Halcon in 1969, is a coproduct process that has been exploited by Oxirane worldwide. The Oxirane process has advantages over the chlorohydrin process becaus it uses oxygen as the oxidizing agent and minimizes by-products.

The Oxirane process is an indirect air-oxidation process utilizing oxidation of isobutane or ethylbenzene to the corresponding hydroperoxide, followed by reaction of the hydroperoxide with propylene to yield propylene oxide and tert-butanol or α-phenethyl alcohol, respectively. The latter is readily dehydrated to styrene. For 1 ton of propylene oxide, either 2.5–3 tons of tert-butanol or 2.8 tons of styrene is produced. The following equations summarize both variations of the Oxirane process.

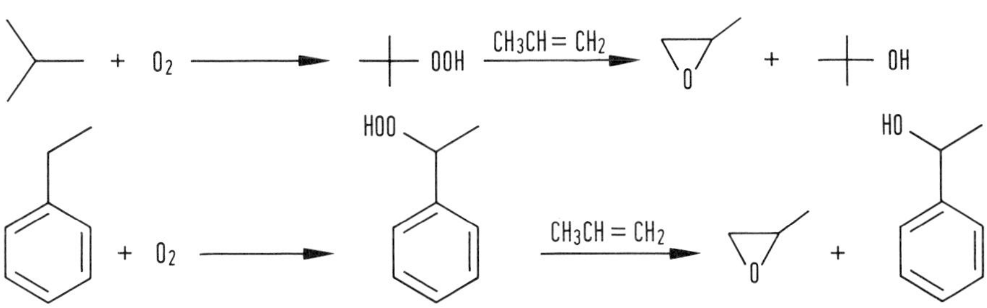

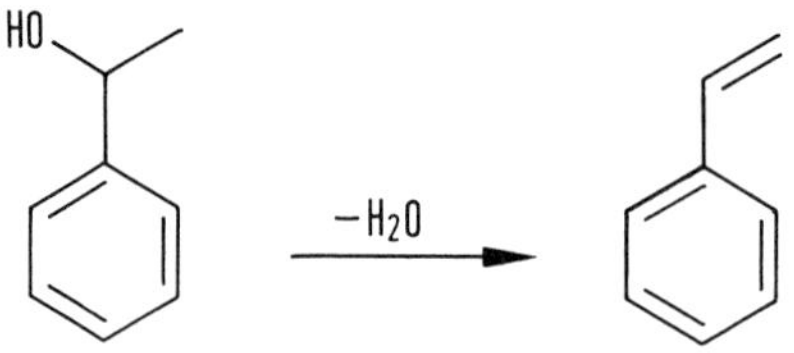

Isobutane is reacted with air at 120–140°C and 25–35 bar to give 48% hydroperoxide in a selectivity of 50%. Reaction of the hydroperoxide occurs at 160–180°C with a selectivity of about 80%. The propylene consumption is 9% (98% selectivity of PO).

The utilization of tert-butanol in a 1:1 mixture with methanol as a gasoline additive by Arco, the present owner of Oxirane, renders the tert-butanol coproduct more useful. The distribution of styrene monomer vs. tert-butanol coproduct in today's Oxirane plants is shown in Table 22.

Table 22. Product Distribution in the Oxirane Coproduct Propylene Oxide Plants

Plant Location	PO Capacity (1,000 tons/yr)	Coproduct
Bayport, Texas	398,000	TBA
Channelview, Texas	180,000	SM
Rotterdam, Netherlands	155,000	TBA
Puertollano, Spain	32,000	SM
Chiba, Japan	90,000	SM

In order to have an economically viable situation, the coproduct has to have a large market potential. In the case of the Oxirane process, both coproducts, tert-butanol and styrene monomer, have huge markets of their own.

A third route to propylene oxide is explored by Interox-Carbochimique, Bayer, and Degussa. In this process, perpropionic acid is used to oxidize propylene to propylene oxide.

$$CH_3CH_2COOH + H_2O_2 \longrightarrow CH_3CH_2CO_3H$$

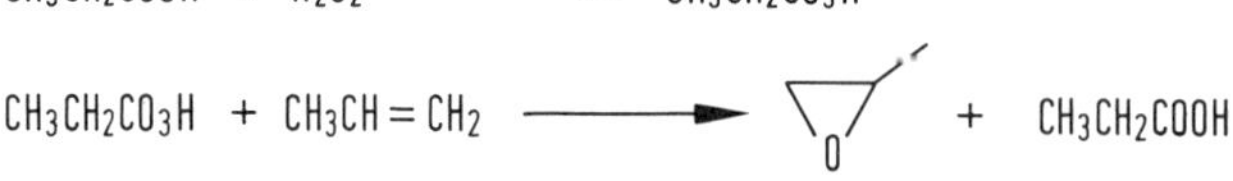

PCUK in France has developed a method for direct oxidation of propylene with hydrogen peroxide, but this process has not yet been commercialized. Daicel in Japan is producing propylene oxide via peracetic acid obtained by air oxidation of acetaldehyde.

Acetoxylation of propylene has been explored by Chem Systems. In this approach propylene is converted to propylene glycol monoacetate, which is thermolyzed to propylene oxide and acetic acid.

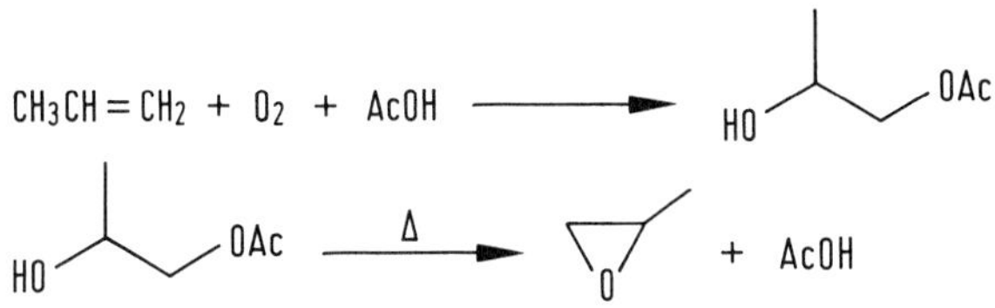

Propylene oxide is used as a monomer for the production of polyether polyols that are used in the polyurethane industry. Especially high molecular weight triols are produced for the manufacture of flexible polyurethane foams (see also under Ethylene Oxide). Propylene glycol, the hydrolysis product of propylene oxide, is used as antifreeze and brake fluid and in the manufacture of alkyd and polyester resins.

4. Acetone CH_3COCH_3

Consumption (1983):	United States	851,000 tons
	Western Europe	784,000 tons
	Japan	346,000 tons
Production Process:	Coproduct in the cumene oxidation to phenol (70%); dehydrogenation of isopropanol; oxidation of propylene	
Major Uses:	Methyl methacrylate (27%), solvent (22%), methyl isobutyl ketone (12%), bisphenol A (7%)	
Price:	$ 0.55/kg	

By far the most widely used process for the manufacture of acetone is the Hock phenol process (cumene process). Air oxidation of cumene produces phenol and the coproduct acetone.

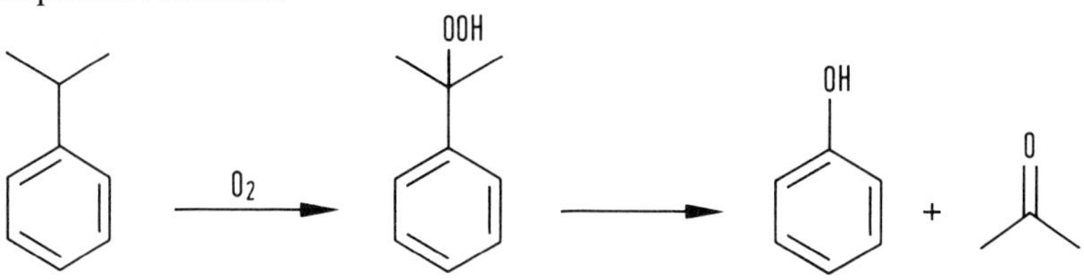

Variations of this process are used for the manufacture of resorcinol and hydroquinone, and in these reactions acetone is also generated as a coproduct.

Dehydrogenation of isopropanol is the second major process for the manufacture of acetone. This reaction is conducted in the vapor phase using ZnO as the catalyst.

Oxidative vapor-phase dehydrogenation has also been used. Shell and Dupont did use the liquid-phase oxidation of isopropanol to manufacture hydrogen peroxide.

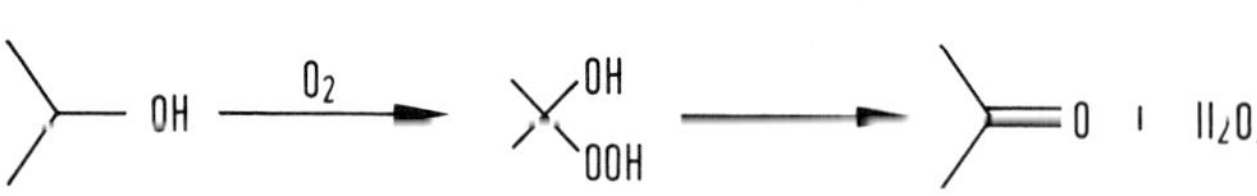

Hoechst has adapted the Wacker process to achieve the direct oxidation of propylene to acetone. The catalyst system is $PdCl_2/CuCl_2$, and the oxidation is conducted in the liquid phase at 110–120°C and 10–14 bar.

$$CH_3CH{=}CH_2 + 0.5\ O_2 \longrightarrow CH_3COCH_3$$

Secondary products derived from acetone include methyl isobutyl ketone (MIBK), methyl methacrylate, and bisphenol A. The major products obtained by aldol condensation of acetone are excellent solvents for the paint industry.

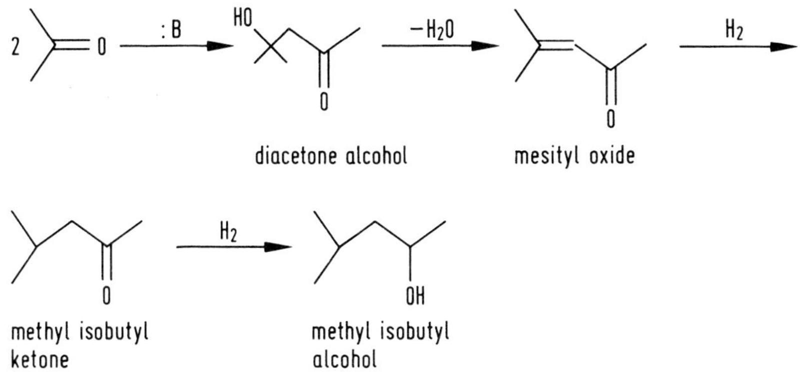

If the base-catalyzed aldol condensation is conducted under more severe conditions, isophorone is obtained. Veba Chemie has utilized isophorone to produce dicarboxylic acids, diamines, and diisocyanates. The difunctional monomer are used in the production of polyamides, polyester, and polyurethanes.

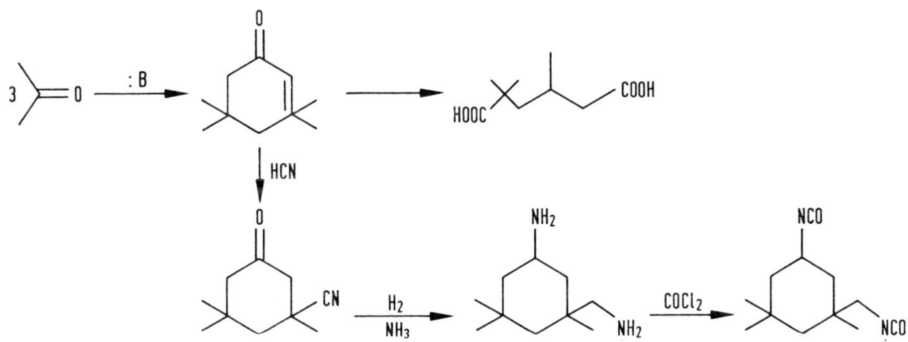

Methacrylic acid and esters are also obtained from acetone and hydrogen cyanide.

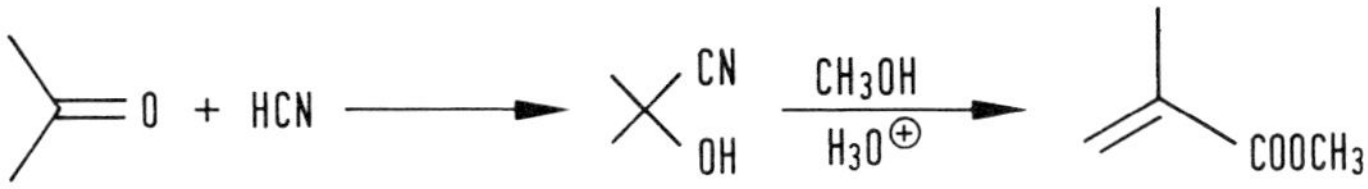

The condensation product of acetone and phenol (bisphenol A) is used as a monomer for polycarbonates and epoxy resins.

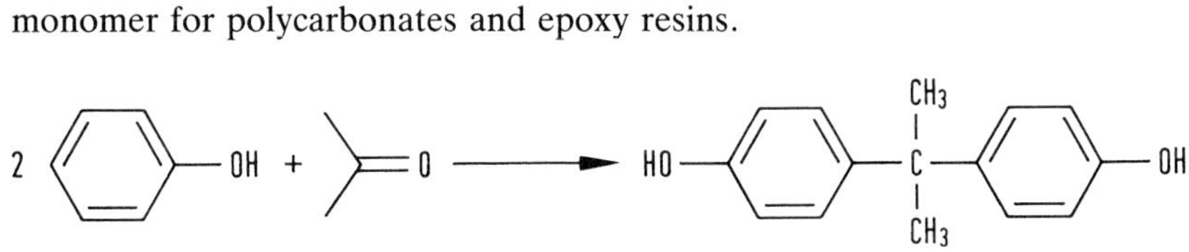

The United States consumption of polycarbonates and epoxy resins in 1984 was 120,000 and 186,000 tons, respectively.

5. Acrylic Acid and Esters

Consumption (1980):	United States	278,000 tons
	Western Europe	200,000 tons
	Japan	83,000 tons
Production Process:	Oxidation of propylene; reaction of ethylene oxide with hydrogen cyanide	
Major Uses (Esters):	Coatings (45%), textile adhesives and binders (24%), fibers (5%)	
Price (Acrylic acid):	$ 1.50/kg	

The classical Reppe process for the manufacture of acrylic acid based on the carbonylation of acetylene has been phased out in the United States. Processes based on propiolactone and on acrylonitrile as feedstocks were abandoned prior to 1978. Today acrylic acid is manufactured in the United States exclusively by air oxidation of propylene involving acrolein as an intermediate.

$$CH_2CH{=}CH_2 \xrightarrow{O_2} CH_2{=}CHCHO \xrightarrow{O_2} CH_2{=}CHCOOH$$

UCC introduced the continuous two-step oxidation process in 1970. Another process based on ethylene oxide is still being used in Europe. In this process ethylene oxide reacts with hydrogen cyanide to give hydroxypropionitrile, which is dehydrated and hydrolyzed to acrylic acid.

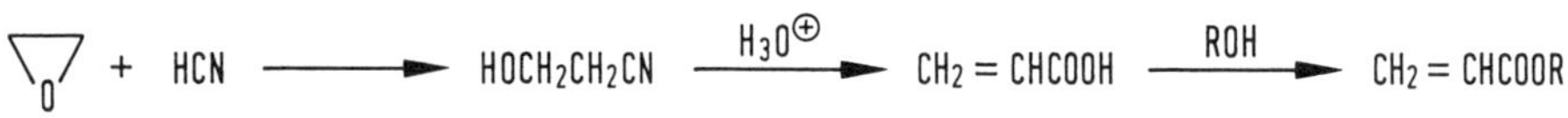

The acrylic acid is converted to its esters by reaction with the appropriate alcohol. The breakdown of acrylate ester production in the United States in 1979 is as follows: ethyl (40%), butyl (39%), 2-ethylhexyl (9%), methyl (7%), and others (5%). Acrylic acid polymers constitute over 85% of the consumption of water-soluble resins, usually in the form of their sodium salts. Copolymers of acrylic acid are also manufactured; typical comonomers are acrylamide and ethylene.

6. Acrylonitrile $CH_2=CHCN$

Consumption (1984):	United States	1,045,000 tons
	Western Europe	1,018,000 tons
	Japan	546,000 tons
Production Process:	Ammoxidation of propylene	
Major Uses:	Acrylic fibers (48%), ABS/SAN (22%)	
Price:	\$ 0.83/kg	

In the traditional route to acrylonitrile, inexpensive C_2 units were converted with hydrocyanic acid into acrylonitrile. A typical process developed by UCC and Cyanamid involved the reaction of ethylene oxide with HCN to ethylene cyanohydrin. Dehydration in the liquid or gas phase over alumina gave acrylonitrile.

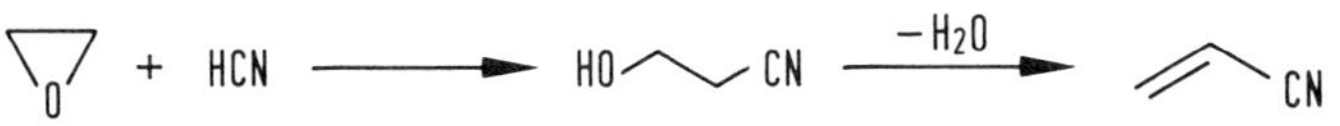

Today acrylonitrile is manufactured almost exclusively by ammoxidation of propylene.

$$CH_2=CHCH_3 + NH_3 + 0.5\ O_2 \longrightarrow CH_2=CHCN + 3\ H_2O$$

In this so-called Sohio process, $Bi_2O_3 \cdot MoO_3$ or $UO_2 \cdot Sb_2O_3$ catalysts are used in a typical vapor-phase process (450°C and 1.5 bar). The selectivity is approximately 67% based on complete conversion of propylene. Propane can also be used as feedstock, but the selectivities are only on the order of 30%. Major United States producers are Dupont, Monsanto, Sohio, and Cyanamid.

Applications of acrylonitrile are mainly the manufacture of acrylic fibers, plastics (ABS, SAN), and nitrile rubber (NBR). Partial hydrolysis of acrylonitrile to acrylamide is also conducted on a commercial scale.

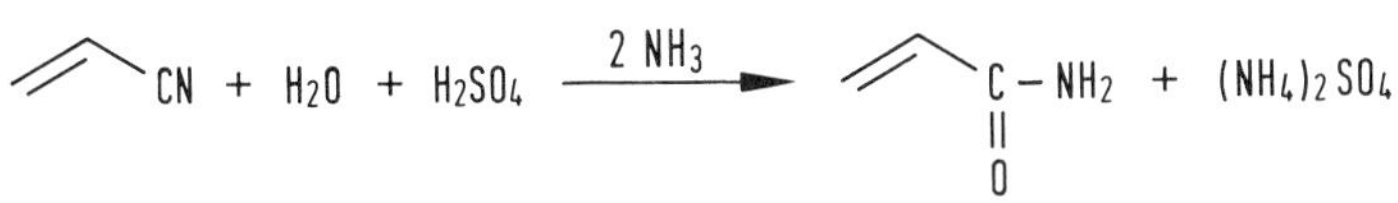

Recently developed processes (Cyanamid, Dow, Mitsui Toatsu, Mitsubishi) avoid sulfuric acid, and the reaction is conducted with water in the presence of a copper-containing catalyst. Acrylamide is used as a monomer for polyacrylamide, which is used in water purification, paper processing, and paints. Polyacrylamide is one of the few water-soluble vinyl polymers.

A recently developed nylon product, nylon-4,6, uses acrylonitrile as the feedstock. Reaction of acrylonitrile with hydrogen cyanide produces succinonitrile, which upon hydrogenation gives 1,4-diaminobutane. Reaction of 1,4-diaminobutane with adipic acid yields nylon-4,6.

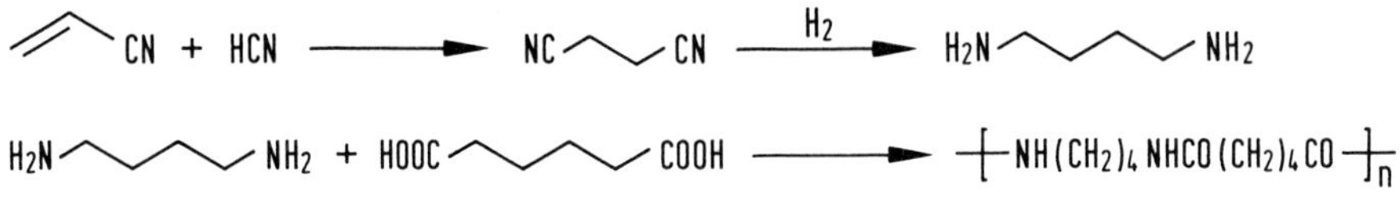

DSM has constructed a pilot plant in Holland to produce Nylon-4,6, which will be marketed under the trade name Stanyl. The product is intended for tire cord and engineering plastics applications.

7. *n*-Butanol $CH_3CH_2CH_2CH_2OH$

Consumption (1983):	United States	364,000 tons
	Western Europe	332,000 tons
	Japan	94,000 tons

Production Process: Hydroformylation of propylene (oxo process) followed by reduction; hydration of l-butene

Major Uses: Butyl acrylate (30%), solvent (18%), butoxyethanol (20%), butyl acetate (13%)

Price: $ 0.68/kg

Reaction of propylene with syngas (hydroformylation) in the liquid phase (140–180°C and 250–300 bar), in the presence of cobalt or rhodium catalysts, produces mixtures of normal and isobutyraldehyde. Reduction of the aldehydes with hydrogen yields *n*-butanol and isobutanol. The ratio of the normal to the iso products is catalyst dependent, but it is usually about 80:20. The UCC low-pressure oxo process uses rhodium phosphine complexes. The rhodium catalyst can be used at lower temperature (100°C), the reaction rate is faster, and the ratio of normal to isobutyraldehyde is 8:1 to 16:1.

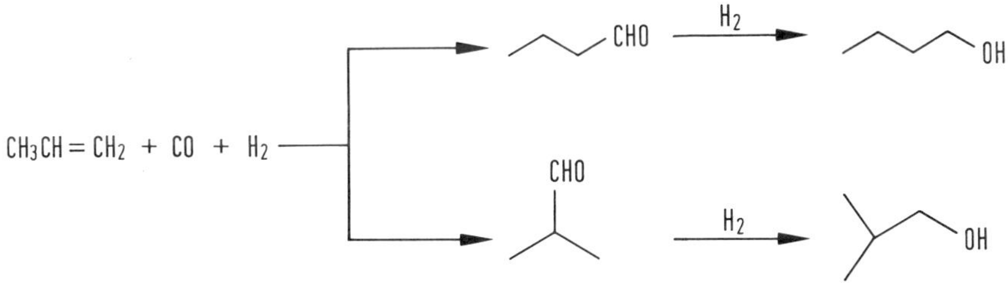

n-Butyraldehyde is the more valuable product because it can be easily converted into *n*-butanol, which is a superior to isobutanol as a solvent and is also the precursor to 2-ethylhexanol. The advantages of the low-pressure oxo process over the high-pressure process are summarized in Table 23.

Table 23. Comparison of High- and Low-Pressure Oxo Processes

	High-Pressure Process	Low-Pressure Process
Pressure (atm)	250–350	20
Temperature (°C)	140–180	100
Normal/iso ratio	3.5:1	10:1
Propylene consumption[a]	965	675
Syngas consumption[b]	1 100	710

[a] Kilograms per 1000 kg *n*-butyraldehyde.
[b] Cubic nanometers per 1000 kg *n*-butyraldehyde.

Shell uses a cobalt trialkylphosphine catalyst (160°C and 35 bar) that directly yields mixtures of the isomeric butanols and 20–30% 2-ethylhexanol.

Reaction of propylene with carbon monoxide and water produces an 85:15 mixture of normal and isobutanol. In this hydrocarbonylation reaction (Reppe process), an iron hydrocarbonyl complex is used as catalyst. This reaction is conducted at 90–110°C and 10–15 bar. The Reppe process is currently used only in Japan.

Secondary and tert-butanol can also be produced by indirect hydration of butenes. From a mixture of 1-butene and 2-butene, only isobutanol is obtained.

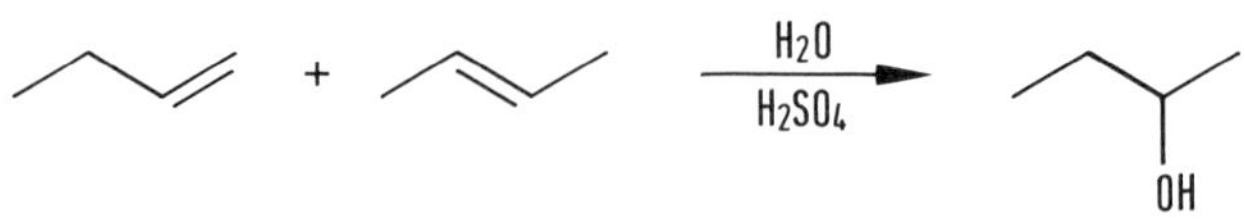

The by-product of the oxo process is isobutanol, and the consumption of this chemical intermediate in the major world markets is summarized in Table 24.

Table 24. Consumption of Isobutanol in the Major World Markets, 1983 (tons)

United States	60,000
Western Europe	220,000
Japan	29,000

By far the most important reaction product derived from the oxo process is 2-ethylhexanol. Aldol condensation of *n*-butyraldehyde gives 2-ethylhexenal, which upon hydrogenation produces 2-ethylhexanol.

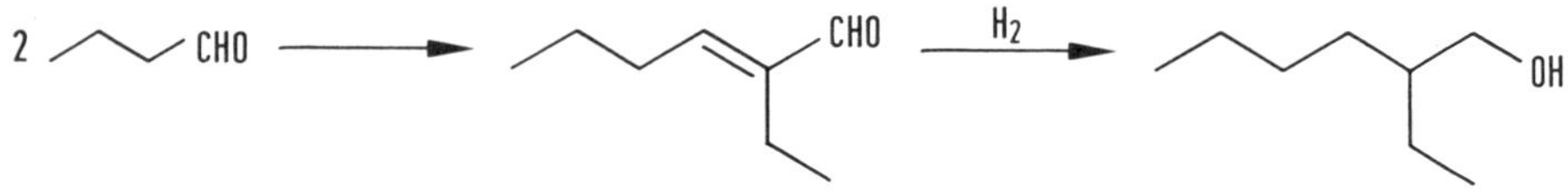

2-Ethylhexanol is the highest volume plasticizer alcohol; it is used to produce the major plasticizer dioctylphthalate. The consumption of 2-ethylhexanol in the major world markets in 1983 is shown in Table 25.

Table 25. Consumption of 2-Ethylhexanol in the Major World Markets, 1983 (tons)

United States	204,000
Western Europe	425,000
Japan	256,000

Another important industrial organic chemical derived from *n*-butyraldehyde is trimethylolpropane. Reaction of formaldehyde with *n*-butyraldehyde in the presence of $Ca(OH)_2$ or NaOH (aldol condensation) followed by reduction produces this polyurethane intermediate.

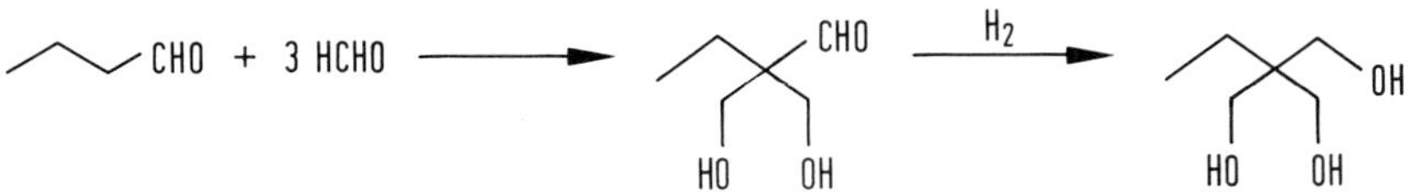

Neopentylglycol, another intermediate for polyesters, polyols, and polyurethanes, is obtained by aldol condensation of isobutyraldehyde and formaldehyde followed by catalytic hydrogenation.

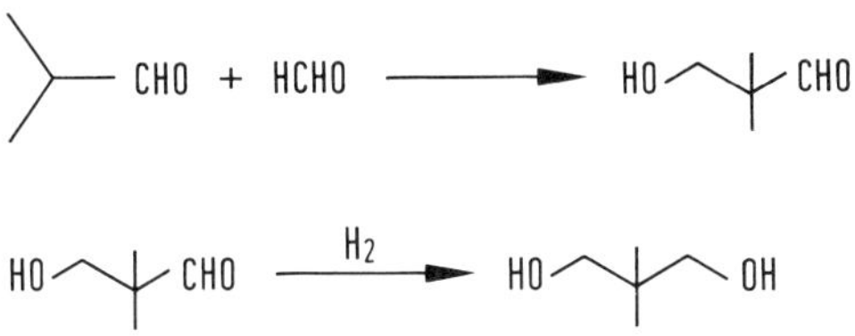

8. Glycerine $HOCH_2CH(OH)CH_2OH$

Consumption (1984):	United States	146,000 tons
	Western Europe	150,000 tons
	Japan	41,000 tons

Production Process: By-product of fatty acid production (53%); allyl chloride via epichlorohydrin (25%); allyl alcohol peracetic acid route (10%)

Major Uses: Drugs and cosmetics (24%), alkyd resins (17%)

Price: $ 1.25/kg

The synthetic routes to glycerine are based on propylene. Hot chlorination of propylene produces allyl chloride, which upon reaction with HOCl affords dichloroisopropanol (see page 102). Treatment with base yields epichlorohydrin. Further hydrolysis leads to glycerine.

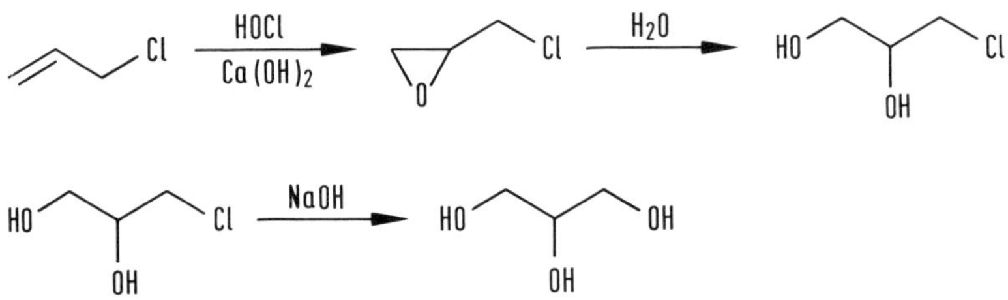

The acrolein/allyl alcohol route used by Shell was abandoned in 1980. In this process air oxidation of propylene gives acrolein, which is reduced with hydrogen to yield allyl alcohol. Reaction of allyl alcohol with hydrogen peroxide then gives rise to the formation of glycerine.

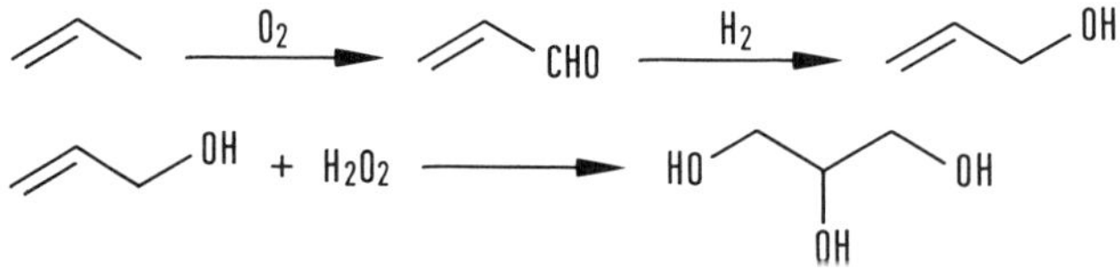

The third route to glycerine involves the rearrangement of propylene oxide to allyl alcohol, followed by peracid oxidation to glycidol. Glycerine is obtained by hydrolysis of glycidol.

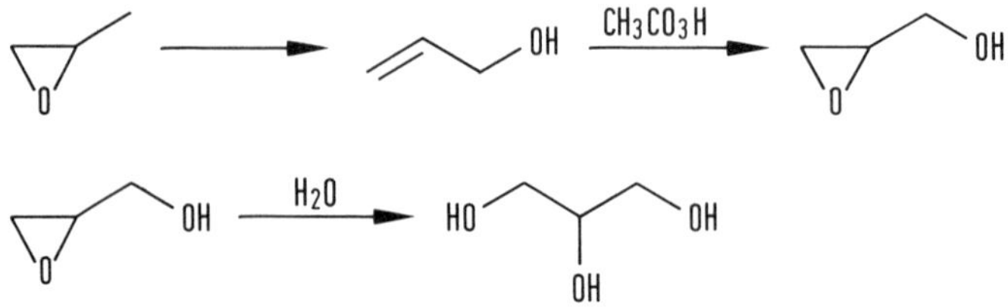

Natural glycerine is produced by hydrolyzing fats or oils with caustic soda.

Glycerine is used mainly in the production of alkyd resins and polyurethane polyols. Drug, food, and cosmetic products are also formulated from glycerine. Peacetime consumption of nitroglycerine, the main ingredient in dynamite, is significant.

9. Epichlorohydrin

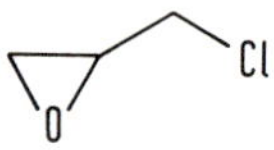

Consumption (1982):	United States	141,000 tons
	Western Europe	141,000 tons
	Japan	36,000 tons
Production Process:	Chlorohydroxylation of allyl chloride	
Major Uses:	Epoxy resins (79%), glycerine (18%), elastomers (3%)	
Price:	$ 1.80/kg	

The manufacture of epichlorohydrin resembles that of propylene oxide. Allyl chloride reacts with hypochlorous acid to give dichloroisopropanol. Treatment of the latter with calcium hydroxide yields epichlorohydrin.

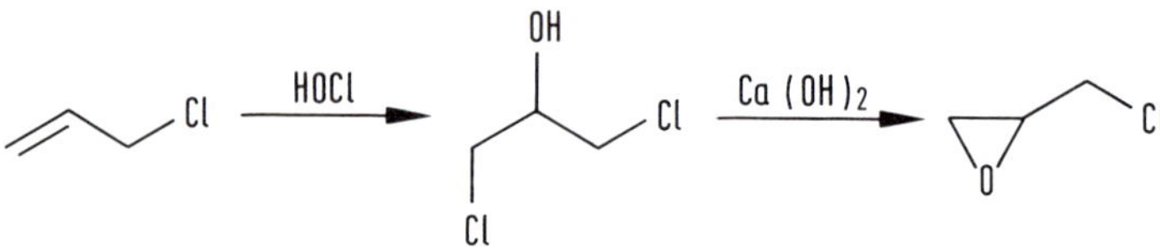

Epichlorohydrin is used in the manufacture of epoxy resins. It is also homopolymerized or copolymerized with ethylene oxide to give epichlorohydrin elastomers with outstanding resistance to ozone, chemicals, and solvents. These elastomers retain useful mechanical properties over a wide temperature range. Their uses include hoses, belting, and gaskets for the automotive industry. The sole United States producer of epichlorohydrin elastomers is Hercules Inc., which has licensed certain types to B. F. Goodrich.

Commercial epoxy resins are based on bisphenol A, which upon reaction with epichlorohydrin produces bisglycidyl ethers. Curing of the glycidyl ethers with di- or polyfunctional molecules containing active hydrogen such as polyamines, polyacids, polymercaptans, and polyphenols produces hard and tough thermoset network polymers with exellent adhesion to many substrates. Epoxy resins have found a wide range of applications, mainly because of their versatility. About half of the epoxy resins are used in protective coatings. The total world consumption of epoxy resins in 1983 was 327,000 tons.

Chapter V. C_4 Chemicals

1. Introduction

The shift to liquid cracking in the United States will produce increased amounts of C_4 feedstock in the years to come. Butane is used to manufacture ethylene and more recently maleic anhydride. Isobutane is an important feedstock for Arco's coproduct propylene oxide process. In this process isobutane is simultaneously oxidized to tert-butanol (see p. 89). The use of tert-butanol as a gasoline additive has prompted development of new processes to convert isobutane to tert-butanol. The C_4 olefins, the butenes and butadiene, account for 60% of the total C_4 feedstock. Major uses of the three butene isomers in the United States are listed in Table 26.

Table 26. Major Uses of Butenes in the United States, 1981 (tons)

Butene-1	58,500	LLDPE (50%), HDPE (42%)
Butene-2	672,600	Butadiene (54%), sec-butanol (45%)
Isobutylene	850,000	MTBE (40%), polybutenes (29%), butyl rubber (17%)

Butadiene is produced mainly by steam cracking of gas oil or naphtha. The growth potential of the butenes as chemical feedstocks is vastly improved by the recent development of LLDPE (linear low-density polyethylene) in which up to 15% of butene-1 is used as comonomer. An exception is Dow's LLDPE process, which uses octene as the comonomer. Another important secondary product is MTBE (methyl *tert*-butyl ether), which is used as a gasoline additive to improve the octane rating. It has been estimated that the demand for C_4 chemicals will reach 7,700,000 tons by the end of this decade. Monomer grade butene-1 is obtained by fractional distillation after removal of isobutylene with sulfuric acid.

2. Butylenes

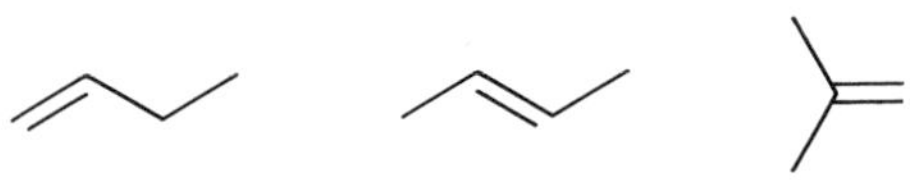

Consumption (1984):	United States	13,568,000 tons
	Western Europe	4,515,000 tons
	Japan	1,423,000 tons
Production Process:	Catalytic cracking and thermal operations; by-product from ethylene production; dehydration of tert-butanol	
Major Uses:	Butadiene, MTBE, polybutylene, sec-butanol, butyl rubber, comonomer for HDPE and LLDPE	
Price:	\$ 0.25–0.40/kg	

The major sources of butylene are refinery off gases from catalytic cracking. The change to zeolite catalysts in high-rise crackers has increased butylene yields. The separation of the butenes can be accomplished by sulfuric acid absorption (to separate isobutylene), *n*-butene isomerization, molecular sieve separation, and simple distillation.

Butadiene is the most important secondary product produced from the C_4 feedstock. It is obtained by the dehydrogenation of *n*-butenes and as a by-product of ethylene manufacture. All of the butadiene produced from *n*-butenes is obtained by oxidative dehydrogenation. In this reaction a preheated mixture of *n*-butene, compressed air, and steam is passed through a fixed-bed catalytic reactor at 510–565°C. The catalyst is a mixture of oxides of tin, bismuth, and boron with phosphoric acid.

Sec-Butanol is produced from butane/*n*-butene mixtures from which isobutylene has been removed. The-*n*-butenes are absorbed in sulfuric acid, and subsequent hydrolysis of the ester produces *sec*-butanol. Approximately 95% of the *sec*-butanol is used to produce methyl ethyl ketone, a commercial solvent. Primary amyl alcohols are obtained from *n*-butenes via the oxo process. Amyl alcohols are used as solvents and diluents.

Reaction of isobutylene with methanol produces methyl *tert*-butyl ether (MTBE), an octane enhancer for motor fuels. Polymer uses of isobutylene include butyl rubber. 1-Butene is used as a comonomer for high-density polyethylene (HDPE) and linear low-density polyethylene (LLDPE).

3. Butadiene

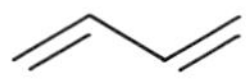

Consumption (1984):	United States	1,227,000 tons
	Western Europe	1,060,000 tons
	Japan	657,000 tons
Production Process:	Steam cracking of gas oil and naphta; dehydrogenation of butylenes or butane	
Major Uses:	Styrene butadiene rubber (45%), polybutadiene rubber (20%), hexamethylenediamine (13%), ABS (7%), polychloroprene (7%)	
Price:	$ 0.48/kg	

The traditional route to butadiene was a stepwise synthesis of the C_4 chain from acetylene or acetaldehyde. For example, aldol condensation of acetaldehyde produces 1,3-butanediol, which is dehydrated in the gas phase at 270°C over a polyphosphate catalyst to give butadiene.

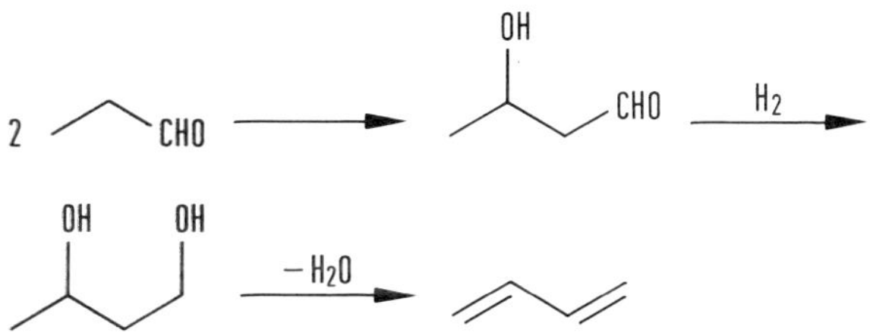

The Reppe process, which is still being used to manufacture 1,4-butanediol (see p. 114), is another classical route to butadiene.

Today butadiene is produced exclusively from petrochemicals. Steam cracking of naphtha or gas oil yields approximately 9% by weight C_4 compounds. The constituents of the C_4 fraction and their boiling points are shown in Table 27. In view of the narrow boiling range, separation by fractional distillation is not possible. Therefore, extractive distillation using cosolvents (DMF, NMP, etc.) is employed.

Table 27. Constituents of the C_4 Fraction from Steam Cracking of Naphtha

	Boiling Point (°C)
Isobutene	−6.90
Butene-1	−6.26
trans-Butene-2	0.88
cis-Butene-2	3.72
1,3-Butadiene	−4.41
Vinylacetylene	5.10

If butane and butene mixtures are available from natural gas, simple dehydrogenation will produce butadiene.

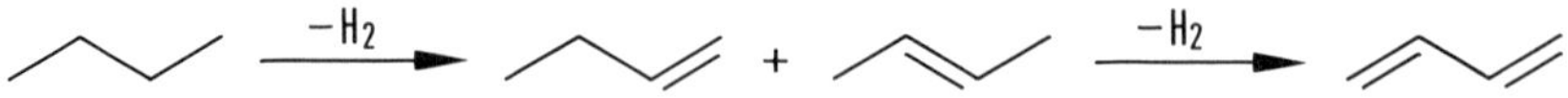

This reaction is usually conducted at 600−700°C with catalysts such as calcium-nickel phosphate stabilized with Cr_2O_3 (Dow process), or $Fe_2O_3 \cdot Cr_2O_3 \cdot K_2O$ (Shell process). These processes have recently been further refined by using air as a coreagent (oxidative dehydrogenation).

Butadiene is used in the production of rubber products (SBR, NBR, polybutadiene). Since these products are used mainly in the transportation industry, butadiene consumption is very dependent on automobile production.

The consumption of butadiene-derived homo- and copolymers in the major industrial markets in 1984 is shown in Table 28.

Table 28. Consumption of Homo- and Copolymers of Butadiene in the Major World Markets, 1984 (tons)

	SBR[a]	NBR[b]	BR[c]	ABS[d]
United States	890,000	51,000	412,000	476,000
Western Europe	654,000	—[e]	235,000	350,000
Japan	302,000	36,000	129,000	368,000

[a] Styrene butadiene rubber.
[b] Nitrile rubber.
[c] Polybutadiene (butadiene rubber).
[d] Acrylonitrile-butadiene-styrene terpolymer.
[e] Not available.

Styrene butadiene rubber (SBR) is the most widely used of the synthetic rubbers, mainly in tire applications. Butadiene rubber (BR) is usually blended with natural rubber or SBR to improve tire tread performance, and nitrile rubber (NBR) is a specialty elastomer with excellent oil resistance.

The largest chemical use for butadiene is in the manufacture of hexamethylenediamine from adiponitrile (see p. 140). Adiponitrile is obtained in the reaction of butadiene with hydrogen cyanide.

Chloroprene, the monomer for polychloroprene (neoprene rubber), is also made from butadiene.

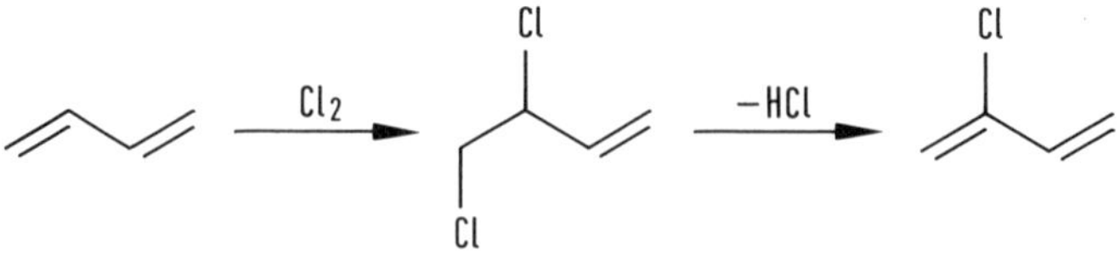

The classical route to chloroprene is based on acetylene. Dimerization of acetylene produces vinylacetylene, which upon addition of hydrogen chloride produces chloroprene.

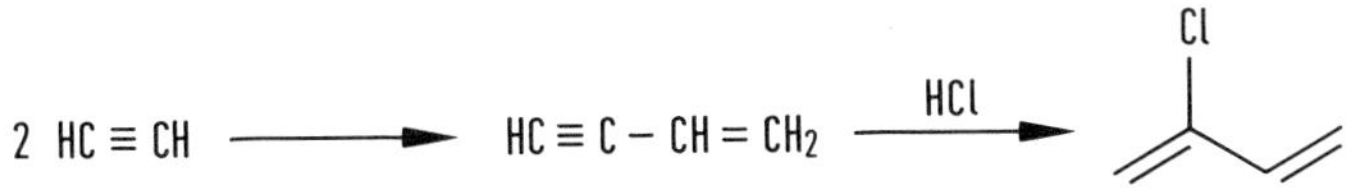

Sulfolane, a high-boiling polar solvent, is manufactured by reacting butadiene with SO_2, followed by hydrogenation.

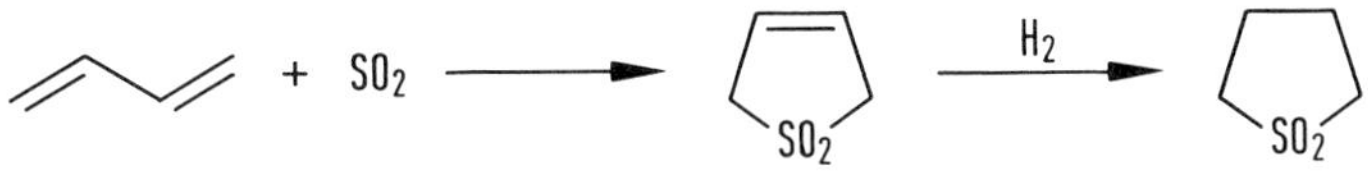

Trimerization of butadiene with Ziegler-Natta type catalysts ($TiCl_4$, $AlCl_2Et$/$AlClEt_2$) was pioneered by Hüls in 1975. By-products in this process are the dimeric products vinyl cyclohexene and cyclooctadiene. The cyclic trimer is the starting material for dodecanedioic acid and laurolactam.

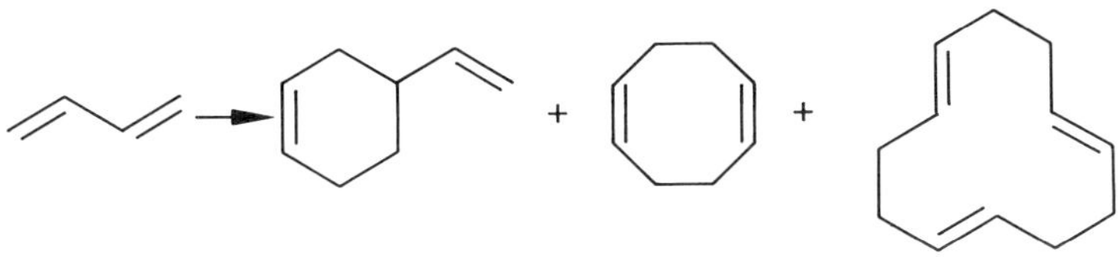

Reaction of cyclooctadiene with ethylene in the presence of a metathesis catalyst (rhenium on alumina) yields 1,5-hexadiene. This process is currently under development by Shell Chimie at Berre near Marseilles.

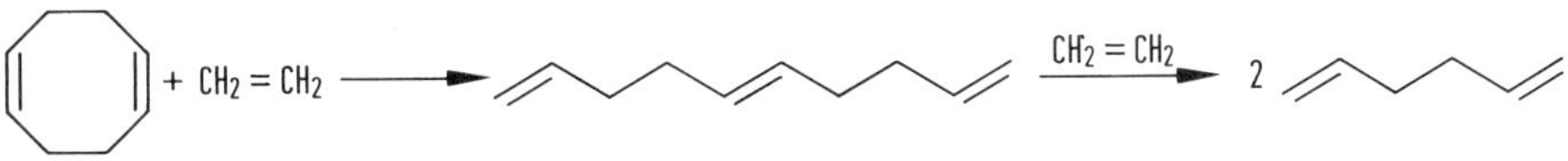

Similar metathesis reactions are used by Shell to produce 1,9-decadiene, 1,5,9-decatriene, and 1,13-tetradecadiene.

The cyclic trimer of butadiene, cyclododecatriene, is hydrogenated over a nickel catalyst (200°C and 10 bar) and oxidized with air to a mixture of alcohol and ketone, very much like the air oxidation of cyclohexane. The alcohol/ketone mixture is further oxidized with nitric acid to give dodecanedioic acid, a comonomer for the manufacture of Qiana fiber. Dehydrogenation of the alcohol/ketone mixture to cyclododecanone, using a copper chromite catalyst, and reaction with hydroxylamine affords the corresponding oxime, which undergoes the Beckmann rearrangement at 100–130°C in oleum to give laurolactam, the monomer for nylon-12. Laurolactam is insoluble in water and separates upon dilution (see Scheme 2). The butadiene dimer 1,5-cyclooctadiene is used as a comonomer in the manufacture of EPDM (ethylene propylene diene monomer) elastomers.

Butadiene is also used as a raw material for the production of 1,4-butanediol by Toyo Soda and Mitsubishi in Japan. While Toyo Soda goes through 1,4-dichlorobutene-2 as an intermediate, Mitsubishi uses acetoxylation technology. The production of adipic acid from butadiene and carbon monoxide is under development. The use of butadiene as a feedstock for adipic acid depends on the relative price difference between cyclohexane and butadiene.

Scheme 2

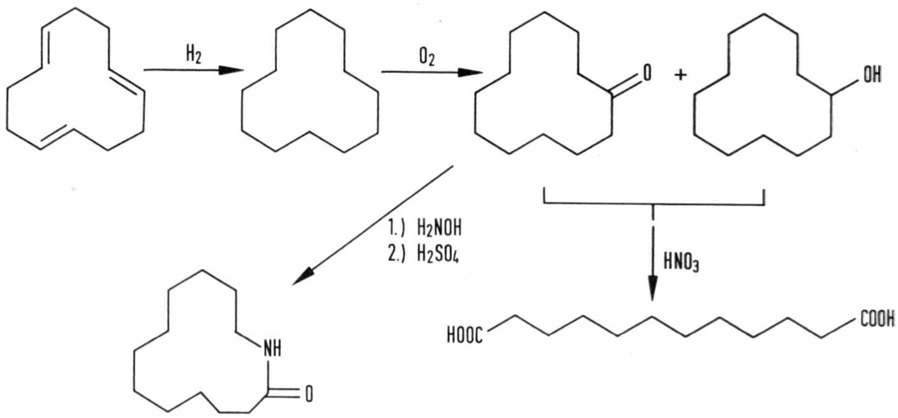

4. Methyl Methacrylate

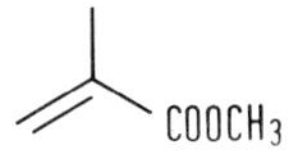

Consumption (1982):	United States	363,000 tons
	Western Europe	235,000 tons
	Japan	179,000 tons
Production Process:	Acetone and hydrogen cyanide	
Emerging Technology:	Oxidation of isobutylene; propylene and syngas.	
Major Uses:	Glazing (28%), surface coating (24%), molding and extrusion (21%)	
Price:	$ 1.36/kg	

The classical route to methacrylic acid and its esters involves the reaction of acetone with hydrogen cyanide to give acetone cyanohydrin. Hydrolysis of this intermediate with concentrated sulfuric acid at 80–140°C affords methacrylic acid amide, which is subsequently converted into methyl methacrylate with methanol. Today all methyl methacrylate produced in the United States is based on the acetone cyanohydrin process.

Intensive research on the use of other feedstocks for methyl methacrylate is under way worldwide. Air oxidation of isobutylene or other C_4 feedstocks containing isobutylene proceeds via methacroleine, with molybdenum as the catalyst to give methacrylic acid.

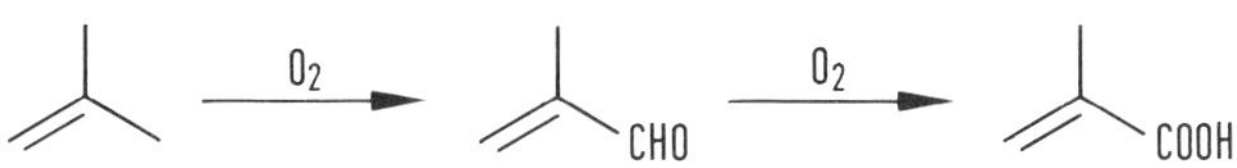

A two-step process starting from *tert*-butanol is also under consideration, where the first step is thermolysis to isobutylene and water.

Other new processes start with propylene or halopropenes. Reaction of propylene with syngas gives isobutyric acid, which can be dehydrogenated to methacrylic acid. 2-Bromopropene reacts with carbon monoxide and methanol to give methyl methacrylate.

Poly(methyl methacrylate) is a clear plastic that is used in glazing and lighting applications. Other major applications include surface coatings, especially for the automotive industry, and injection molding and extrusion. The 2-hydroxyethyl ester of methacrylic acid is the monomer of choice for soft contact lenses. Copolymers of methyl methacrylate with acrylates, acrylonitrile, styrene, butadiene, and other monomers are also produced.

5. Maleic Anhydride

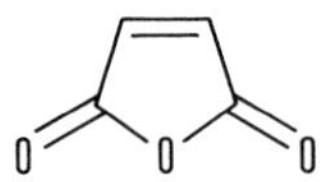

Consumption (1983):	United States 151,000 tons
	Western Europe 120,000 tons
	Japan 45,000 tons
Production Process:	Benzene oxidation; butane und butene oxidation; by-product from phthalic anhydride production
Major Uses:	Unsaturated polyester (69%), agricultural chemicals (12%), lubricating oil additives (12%)
Price:	\$ 1.15/kg

Until recently benzene was the exclusive raw material for maleic anhydride, but due to availability of the less expensive C_4 feedstock, a switch to butane is taking place. After separation of butadien, the C_4 fraction is directly oxidized to maleic anhydride. This reaction is carried out using air and a fixed-bed catalyst based on vanadium, modified with other oxides (tellurium, molybdenum, antimony) at 350–450°C and 2–3 bar.

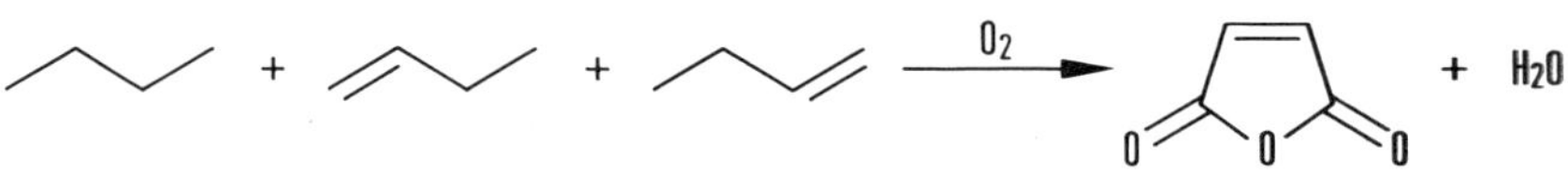

The selectivity to maleic anhydride is usually on the order of 50–60%. Mitsubishi in Japan is using a V_2O_5/H_3PO_4 catalyst that is whirled in a fluidized bed to uniformly remove the heat that is generated. A switch from fixed-bed reactors to fluidized-bed reactors, in which the gaseous feedstock enters the bottom of the reactor and keeps the particles of the catalyst bed continuously in suspension, is anticipated. Sohio, in a joint venture with UCB, has successfully used a fluidized bed reactor at a large pilot facility in the United States. Molar yields well in excess of 50% were achieved on a once-through basis.

Monsanto is building the world's largest maleic anhydride facility (58,000 tons/year) at Pensacola, Florida, using the fixed-bed process. Nearly all of the United States maleic anhydride capacity is now based on butane as the feedstock. It takes 20–25% less butane than benzene to produce a pound of maleic anhydride. In addition, butane is priced at its fuel value, which is about half the benzene price. Production of liquid petroleum gas, which contains about 40% butane, is booming in the Middle East, Africa, Mexico, and Canada, which assures a stable supply.

The oxidation of benzene employs similar catalysts at 400–450°C and 2–5 bar and a residence time of 0.1 sec.

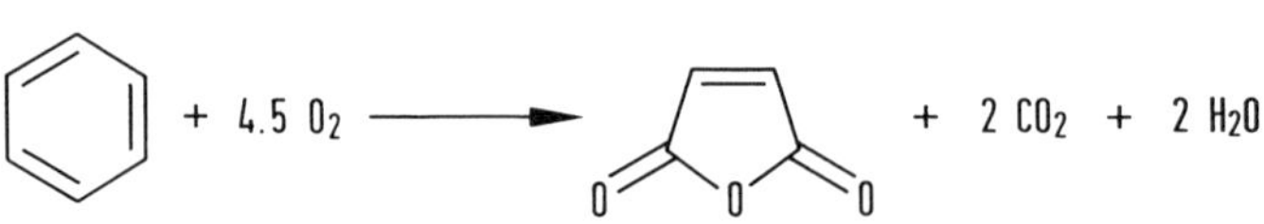

Benzene conversion is on the order of 85–95%, and the selectivity to maleic anhydride is 50–60%.

Maleic anhydride is used mainly in the manufacture of unsaturated polyester resins. Reduction of maleic anhydride to tetrahydrofuran is also of commercial importance. Addition of water to maleic anhydride gives malic acid, which is used as a food flavor additive.

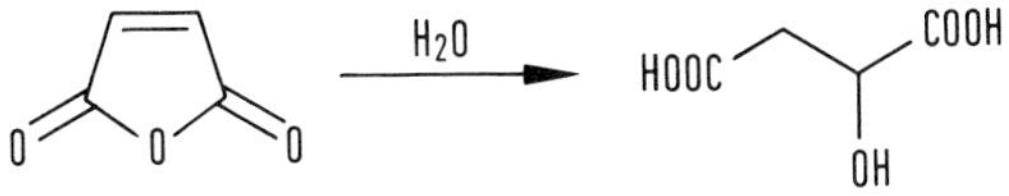

The agricultural chemical applications of maleic anhydride include the insecticide malathion (5 800 tons/year), the plant growth inhibitor maleic hydrazide (2 400 tons/year), and the fungicide captan (2 000 tons/year). This is one of the few examples where basic building blocks are utilized in the manufacture of derivatives with diverse biological activities. The synthesis of captan involves butadiene and maleic anhydride:

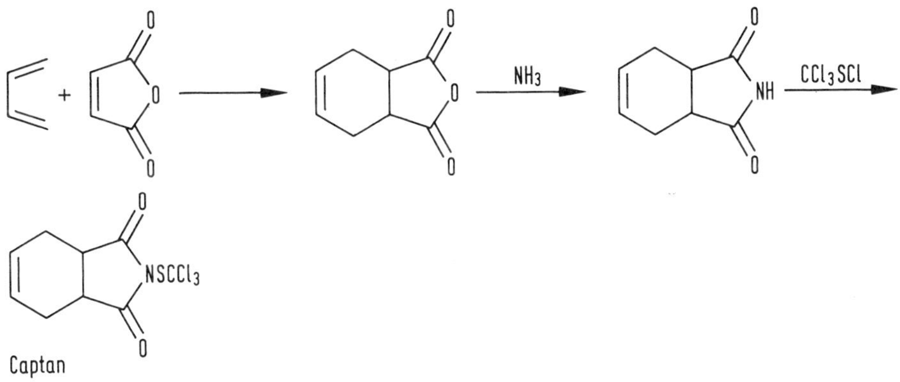

Unsaturated polyester resins are made from maleic anhydride and polyhydric alcohols (glycols); the reaction products are subsequently dissolved in styrene monomer. After compounding with fillers such as fiberglass and calcium carbonate, fabrication is achieved by crosslinking with free radical initiators. The product is generally referred to as SMC (sheet molding compound), and large-volume uses include boats and automotive parts.

Another use for maleic anhydride is in the synthesis of copolymer polyols. Reaction of maleic anhydride with glycols, followed by propoxylation, yields unsaturated polyols, to which styrene and acrylonitrile are grafted. These copolymer polyols are used in the manufacture of highly resilient and flexible polyurethane foams.

6. 1,4-Butanediol 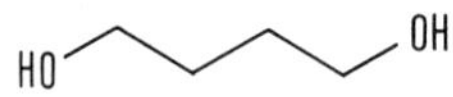

Consumption (1984):	United States	150,000 tons
	Western Europe	100,000 tons
	Japan	20,000 tons
Production Process:	Reaction of acetylene with formaldehyde	
Emering Technology:	Acetoxylation of butadiene; allyl alcohol and carbon monoxide; hydrogenation of maleic acid esters	
Major Uses:	Polyurethanes, polyesters, tetrahydrofuran, PBT	
Price:	$ 1.90/kg	

Butanediol is one of the few chemicals that is still produced from acetylene. Reaction of acetylene with aqueous formaldehyde at 100°C and 5 bar, in the presence of a modified copper acetylide catalyst, gives 2-butyne-1,4-diol. Reduction with hydrogen in the liquid phase at 70–100°C and 250–300 bar in the presence of Raney nickel yields 1,4-butanediol.

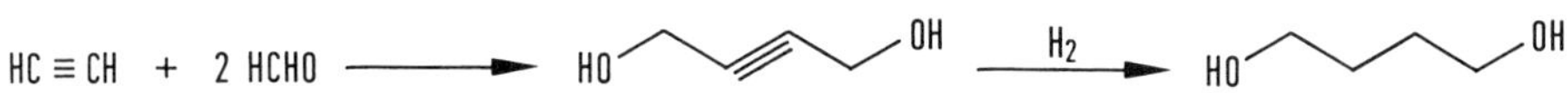

Emerging process technology is based on the acetoxylation of butadiene using a supported palladium catalyst. Also, oxidation of butane to maleic anhydride, followed by hydrogenation of a maleic acid ester, will give 1,4-butanediol.

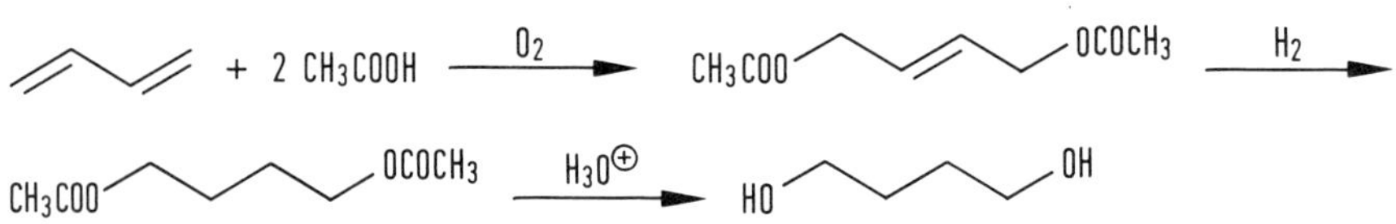

Another attractive route to 1,4-butanediol is the rhodium-catalyzed hydroformylation of allyl alcohol. Allyl alcohol is readily obtained by the acid-catalyzed rearrangement of propylene oxide:

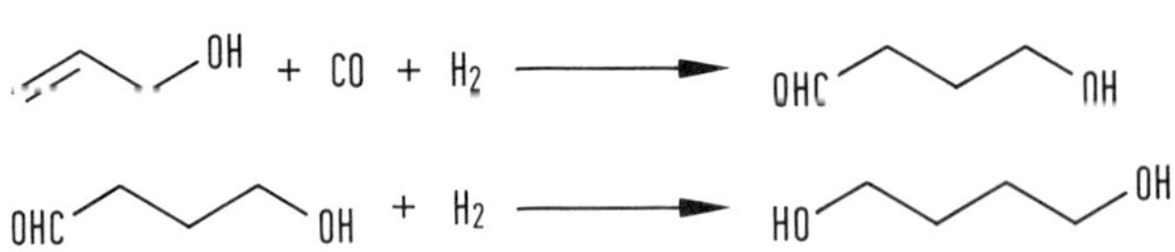

Butanediol is used as a building block for thermoplastic polyurethane elastomers. Also, poly(butylene terephthalate) (PBT), an engineering thermoplastic, uses butanediol as comonomer. γ-Butyrolactone, the starting material for *N*-methylpyrrolidone, is produced by the dehydrogenation of butanediol. Reaction of γ-butyrolactone with ammonia yields 2-pyrrolidone, which upon reaction with acetylene yields *N*-vinyl-2-pyrrolidone.

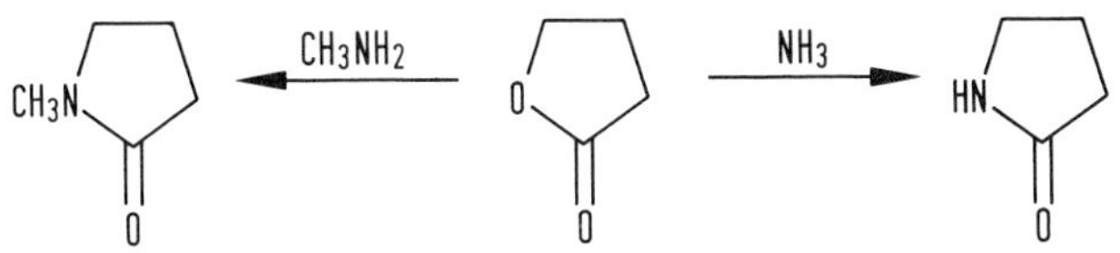

N-Vinyl-2-pyrrolidone is the monomer for polyvinylpyrrolidone as well as styrene/vinylpyrrolidone copolymers.

Dehydration of butanediol gives tetrahydrofuran.

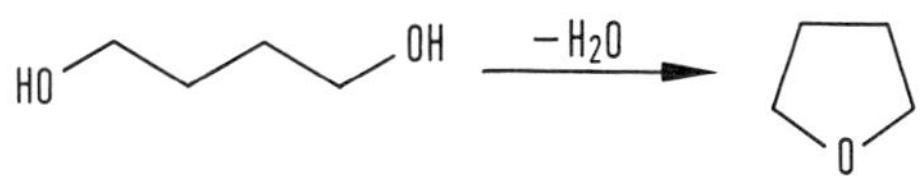

Tetrahydrofuran is also manufactured from furfural, which is obtained from agricultural waste materials containing pentosans (see p. 124).

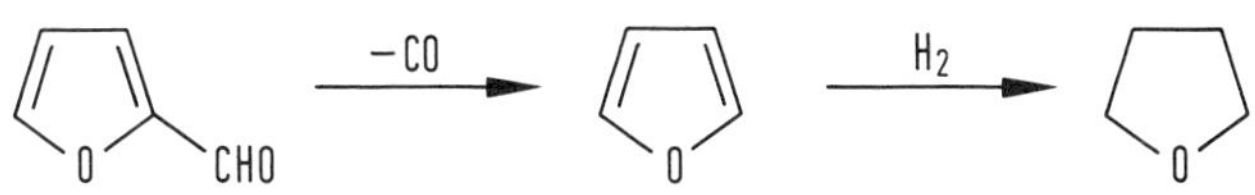

Tetrahydrofuran is the monomer for poly(tetramethylene glycol), a basic building block for segmented polyurethane elastomers and segmented polyester elastomers. The homopolymerization of tetrahydrofuran to poly(tetramethylene glycol) is conducted in the presence of Lewis acid catalysts, such as fluorosulfonic acid or perchloric acid.

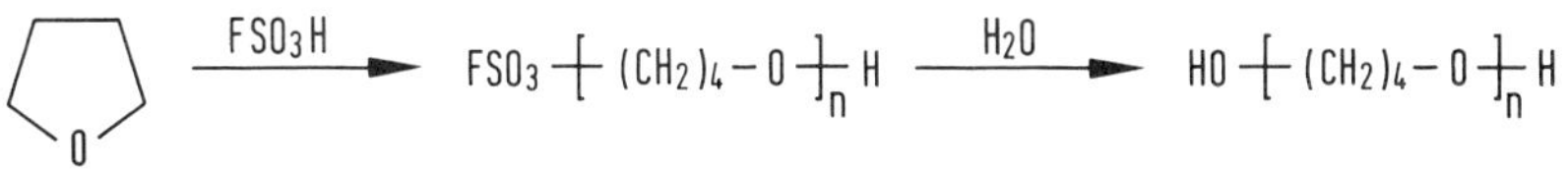

The worldwide consumption of poly(tetramethylene glycol) in 1982 was approximately 26,000 tons.

Tetrahydrofuran is also used as a polar solvent in adhesives and coatings applications.

Chapter VI. C_5 Chemicals

1. Introduction

The trend toward the use of naphtha feedstock also provides ready access to the C_5 cracking fraction. The C_5 fraction contains pentanes, pentenes, cyclopentadiene, isoprene, and piperylene (1-methyl-1,3-butadiene). By far the most important component is isoprene, which is used mainly as a monomer for *cis*-1,4-polyisoprene, the synthetic version of natural rubber. In the United States all of the isoprene monomer is obtained from petrochemical feedstocks, either by direct solvent extraction or by dehydrogenation of isopentane and isopentenes. Cyclopentadiene, a precursor of norbornene, 5-ethylidene norbornene, and hexachloropentadiene, is another important industrial chemical. The separation of cyclopentadiene is easily accomplished by heating the C_5 fraction to 140–150°C. Under these conditions cyclopentadiene dimerizes to dicyclopentadiene, which has a much higher boiling point. Cyclopentadiene can be regenerated from its dimer by heating above 200°C.

Other commercially important C_5 chemicals, such as valeric acid, amyl alcohols, pentaerythritol, and neopentyl glycol, are synthesized from lower carbon chain feedstocks. Furfural is obtained from agricultural waste products containing pentosans.

2. Isoprene

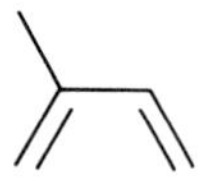

Consumption (1982):	United States	60,000 tons
	Western Europe	48,000 tons
	Japan	73,000 tons
Production Process:	Isolation from C_5 fractions; dehydrogenation of isopentane and isopentenes; dimerization of propylene; from isobutylene and formaldehyde; from acetone and acetylene	
Major Uses:	Polyisoprene (95%), butyl rubber	
Price:	\$ 0.53/kg	

Isoprene is used primarily in synthetic rubber products. Isotactic 1,4-*cis*-polyisoprene is identical to natural rubber Many countries especially in the Comecon countries and in South Africa, are making an effort to become independent of natural rubber. Production of polyisoprene in the Comecon countries is more than double that of the rest of the world. South Africa is building a chemical complex to produce synthetic rubber from coal.

In the United States the production of isoprene is based on petrochemicals. Isoprene can be directly isolated from C_5 fractions obtained in the cracking of naphtha, it can be obtained by dehydrogenation of isopentane and isopentene mixtures, or it can be produced by dimerizing propylene to isohexene using Ziegler-Natta catalysts. Isomerization of 2-methyl-1-pentene to 2-methyl-2-pentene is readily accomplished on a supported acid catalyst. Treatment of 2-methyl-2-pentene with steam in the presence of catalytic amounts of HBr yields isoprene:

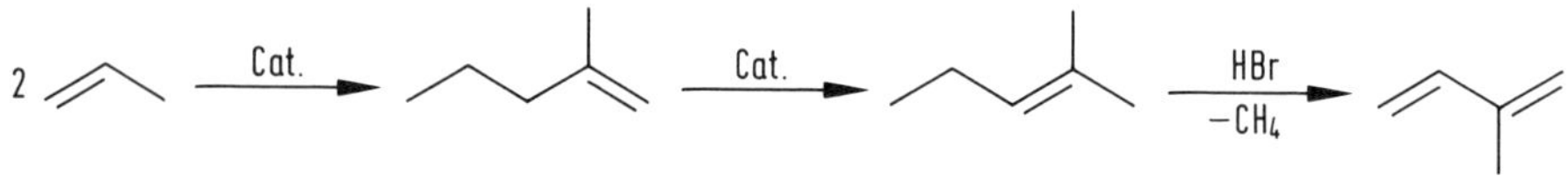

Goodyear operates a 60,000 ton/year plant in the United States that is based on the dimerization of propylene. Another synthetic route to isoprene based on isobutylene and formaldehyde is used in Western Europe, USSR, and Japan. Acid catalysis is used in the formation of the dioxane intermediate as well as in its thermolysis.

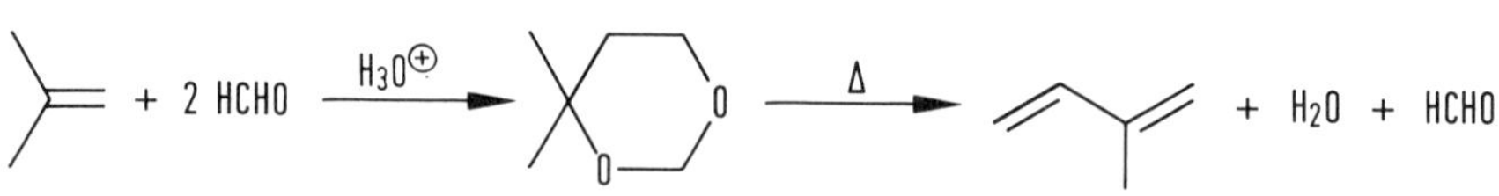

In Italy, Isoprene is synthesized from acetone and acetylene. The first step is conducted in liquid ammonia at 10–40°C and 20 bar. Subsequent hydrogenation and dehydration produces isoprene.

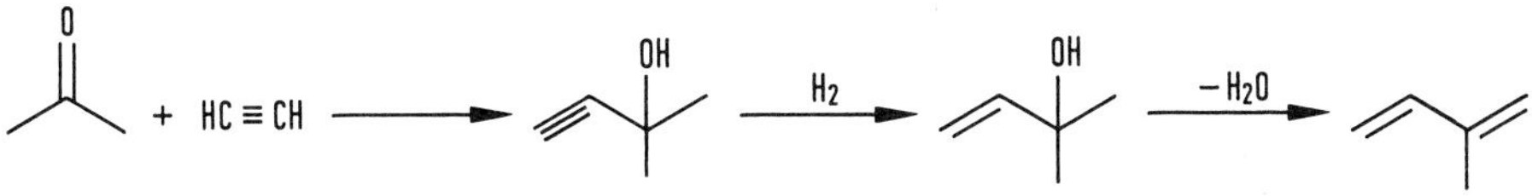

Isoprene is used mainly as a monomer for 1,4-*cis*-polyisoprene and as a comonomer in the production of butyl rubber. The price and availability of natural rubber are important factors in the utilization of polyisoprene. Rapid growth in the production of radial tires increases the demand for natural rubber or polyisoprene. A small amount of isoprene monomer is polymerized into *trans*-1,4-polyisoprene, a material very much like balata or gutta percha rubber. *trans*-1,4-Polyisoprene is used in the manufacture of golf ball covers and in medical applications. *cis*-1,4-Polyisoprene is used in the manufacture of tire and tire products (55%), mechanical goods (21%), footwear (16%), rubber bands and erasers (11%), and numerous other products.

3. Dicyclopentadiene

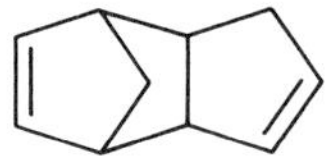

Consumption (1982):	United States 34,500 tons
Production Process:	Isolation from C_5 fractions after heat treatment
Major Uses:	5-Ethylidene norbornene, norbornene, hexachlorocyclopentadiene
Price:	\$ 0.62/kg

Dicyclopentadiene is used as a diene comonomer in the manufacture of EPDM (ethylene propylene diene monomer) elastomers. However, most producers of EPDM elastomers prefer to use 5-ethylidene norbornene as the comonomer. 5-Ethylidene norbornene is obtained from cyclopentadiene and butadiene. The initially formed Diels-Alder adduct is subsequently isomerized to the final product. Cyclopentadiene is readily obtained from dicyclopentadiene by heating above 200°C and distillation of the low-boiling monomer.

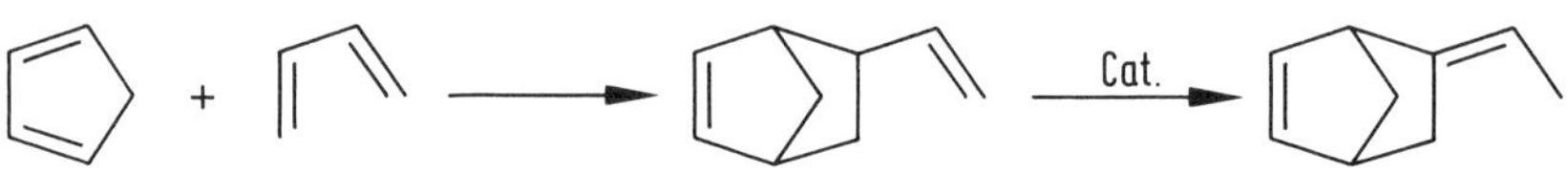

The diene comonomer is used in 3–9% (w/w) portions to provide double bonds in the EPDM elastomer as sites for vulcanization. The total world consumption of EPDM in 1978 was about 270,000 tons. UCC has a plant capacity of 8 200 tons for 5-ethylidene norbornene in the United States.

Reaction of cyclopentadiene with ethylene yields norbornene. Ring-opening polymerization of norbornene produces an amorphous polymer that is used to mop up oil spills.

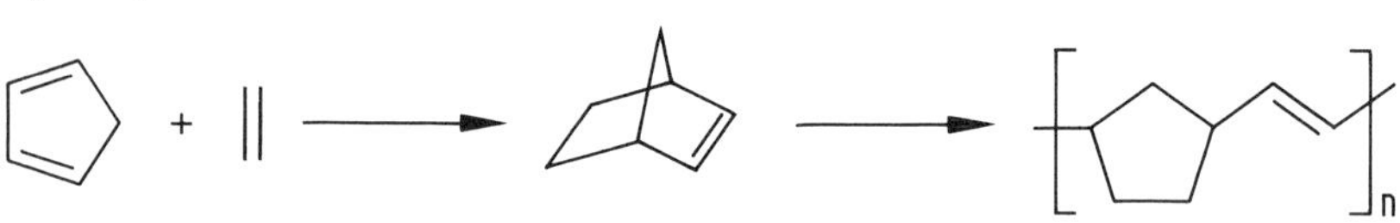

Chlorination of cyclopentadiene gives hexachlorocyclopentadiene, an important starting material for functional fire retardants. Reaction of hexachlorocyclopentadiene with maleic anhydride yields chlorendic anhydride, which is used in the formulation of fire-retardant unsaturated polyester resins. In 1978 approximately 3 200 tons of chlorendic anhydride was used in this application.

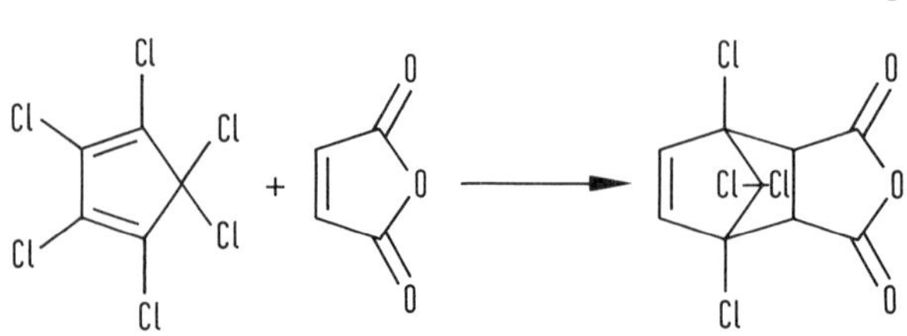

Chlorendate esters are also used as fire retardants in the formulation of rigid polyurethane foams. For example, ring opening with ethylene glycol, followed by ethoxylation or propoxylation gives rise to the formation of a chlorinated polyol.

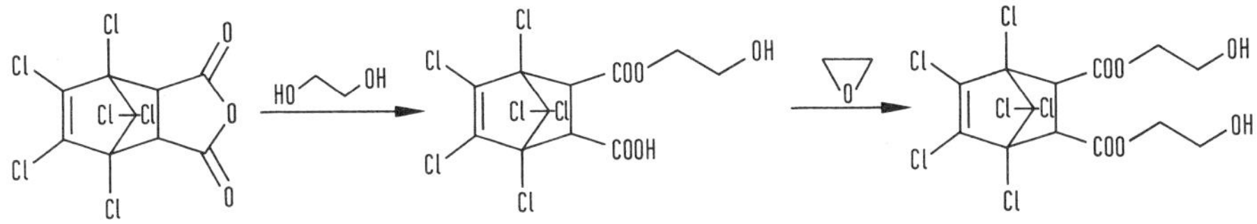

A process for in-mold polymerization of dicyclopentadiene has been developed by Hercules. A liquid stream of dicyclopentadiene monomer is mixed in a RIM (reaction injection molding) machine with a stream of the monomer containing an activator and a retarder. Crosslinked polydicyclopentadiene is formed in the mold. Polydicyclopentadiene is used in the reaction injection molding of snowmobile hoods and other specialty parts.

4. Furfural

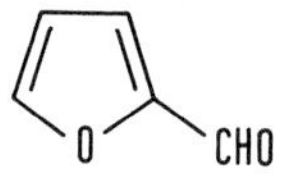

Consumption (1980):	United States	47,000 tons
	Western Europe	59,000 tons
	Japan	9,000 tons
Production Process:	Treatment of agricultural waste (corncobs, sugarcane, bagasse, oat hulls, and rice hulls) with dilute sulfuric acid	
Major Uses:	Furfuryl alcohol (57%), tetrahydrofuran (23%)	
Price:	$ 1.38/kg	

In the United States the Quaker Oats Company is the only producer of furfural (capacity 98,000 tons). Furfural is produced from agricultural waste products containing pentosans. The agricultural waste products are mixed with dilute sulfuric acid, and the pentosans are first hydrolyzed to pentoses, which are subsequently dehydrated to furfural.

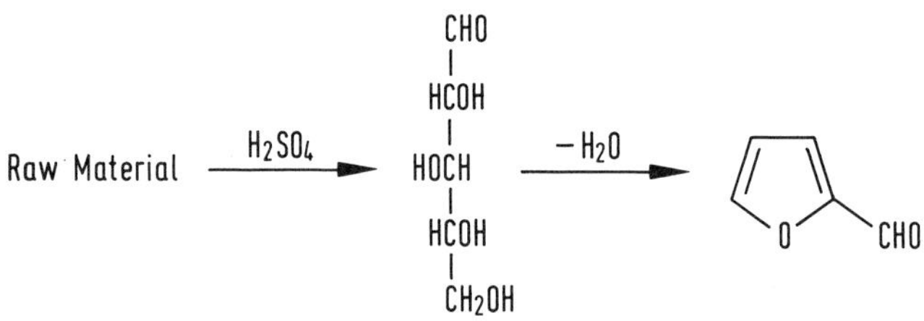

Phosphoric acid or acetic acid can be used in place of the sulfuric acid in the hydrolysis of the raw materials.

The main use of furfural is in the manufacture of furfuryl alcohol. The hydrogenation of furfural to furfuryl alcohol is conducted in the presence of a copper catalyst to assure selectivity.

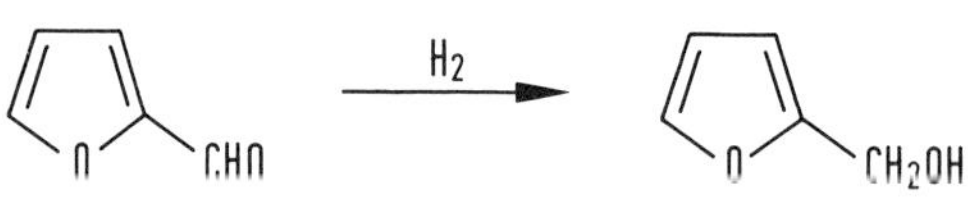

Acid-catalyzed polymerization of furfuryl alcohol produces furan resins, which are used in the bonding of sand for cores and molds in foundry metal casting.

Furfural is also used in the manufacture of tetrahydrofuran, an important monomer and polar solvent. The process of producing tetrahydrofuran from furfural involves decarbonylation to furan followed by catalytic hydrogenation.

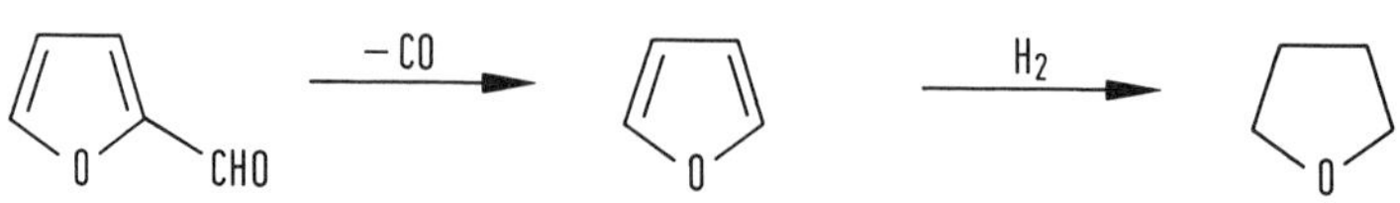

5. Pentaerythritol $HOCH_2C(CH_2OH)_2CH_2OH$

Consumption (1977):	United States	57,000 tons
	Japan	40,000 tons
Production Process:	Aldol condensation of acetaldehyde with formaldehyde	
Major Uses:	Alkyd resins (66%), resins and tall oil esters (16%), fatty acid esters (13%)	
Price:	$ 1.40/kg	

Pentaerythritol is produced by an aldol condensation of acetaldehyde and formaldehyde in the presence of aqueous calcium hydroxide. A mixed aldol condensation takes place, followed by a Canizzaro reaction. The intermediate trimethylol acetaldehyde is reduced by an excess of formaldehyde to give pentaerythritol and formic acid.

$$3\ HCHO + CH_3CHO \longrightarrow HOCH_2C(CH_2OH)_2CHO$$

$$HOCH_2C(CH_2OH)_2CHO \xrightarrow[-HCOOH]{HCHO} HOCH_2C(CH_2OH)_2CH_2OH$$

Pentaerythritol is a building block for unsaturated polyesters as well as polyurethanes. Reaction with epoxides yields compounds useful as surfactants, as reactants in radiation-cured coatings, and as thermoset resins.

Chapter VII. Benzene Derivatives

1. Introduction

The BTX fraction obtained in the catalytic reforming of naphtha is the main raw material source for major aromatic chemicals and polymers. BTX stands for benzene, toluene, and xylenes, and today's synthetic fibers and some major plastic materials are produced from these basic chemicals. Benzene is the raw material for nylon-6,6, nylon-6, polystyrene, polyurethanes, and phenolic resins. Since benzene is used to a larger extent as a chemical raw material, it is not surprising that toluene is converted to benzene by a hydrodealkylation process. Regulatory restrictions on the content of lead in gasoline creates a pressure to use the BTX fraction as an octane booster in gasoline. Refinery conditions determine how much extra benzene is produced by hydrodealkylation of toluene. If the benzene price is weak, hydrodealkylation units shut down. The recent trend in the United States toward the use of naphtha and gas oil as feedstocks for olefin production will generate more benzene by-product. The other benzene source, coal tar, is small, and the decline in steel production in the United States has further reduced the amount of benzene available from the coking of coal.

The basic chemicals derived from benzene are shown in Figure 14.

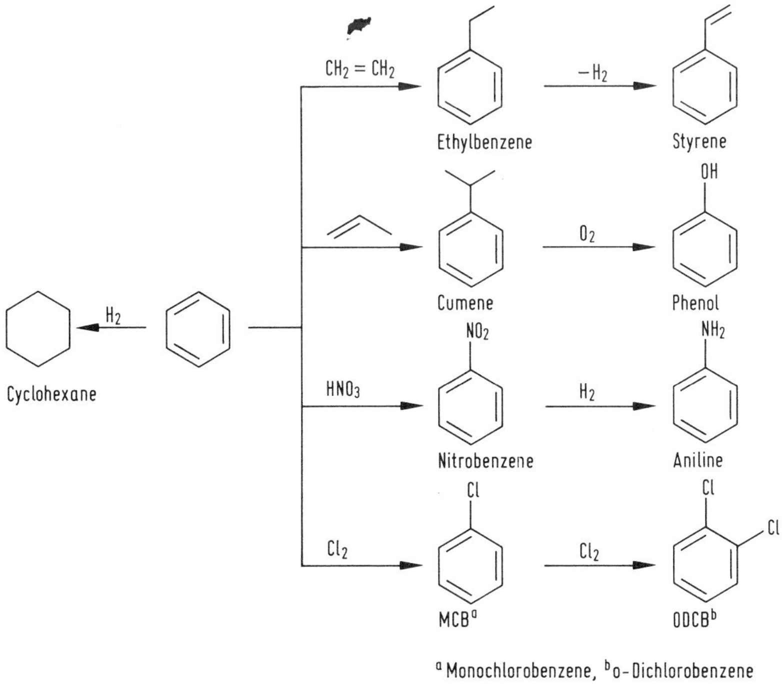

Figure 14. Major products obtained from benzene.

It is interesting to note that the combination of benzene with the other two aliphatic key chemicals, ethylene and propylene, constitutes by far the largest usage for benzene (70%).

Benzene is not only the major raw material for styrene polymers, it is also used in the manufacture of PMDI (polymeric isocyanate based on aniline and formaldehyde), an important building block for polyurethanes. Other important polymers based on benzene include the major engineering plastics (see Table 29) with the exception of polyacetal. Worldwide consumption of these plastics in 1984 was 1.2 million metric tons.

Other high-performance polymers based on benzene include polyarylates, polyether ether ketone (PEEK), polyether imides, and polyphosphates.

Table 29. Consumption of the Major Engineering Thermoplastics, 1984 (tons)

	United States	Western Europe	Japan
Nylons	181,000	235,000	81,000
Polycarbonates	136,000	90,000	42,000
MPPO[a]	77,000	32,000	40,500
Polyacetals	54,000	65,000	65,000
PBT[b]	50,000	14,000	27,000
Polysulfone	6,000	1,200	600
PPS[c]	4,500	900	3,000

[a] Modified poly(phenylene oxide).
[b] Poly(butylene terephthalate).
[c] Poly(phenylene sulfide).

The engineering thermoplastics have growth rates (5–10%) that far exceed the growth rates of the commodity plastics. Some of the smaller volume engineering thermoplastics will even grow at 10–30% per year.

2. Benzene

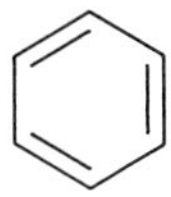

Consumption (1984):	United States	5,532,000 tons
	Western Europe	5,221,000 tons
	Japan	2,102,000 tons
Production Process:	Separation from aromatic fractions in oil refineries; hydrodealkylation of toluene; coking of coal	
Major Uses:	Ethylbenzene (50%), cumene (20%), cyclohexane (15%), aniline (5%)	
Price:	$ 0.35/kg*	

In 1978 about 44% of the total United States benzene production was derived from catalytic reformate (2,237,000 tons). Catalytic reforming is a process in which naphtha is subjected to catalytic dehydrogenation in the presence of a platinum catalyst. The benzene content in reformate streams range from 2 to 10%. Usually half the benzene is separated and the remainder is left in the reformate for gasoline consumption. Benzene is also generated as a by-product in olefin plants (steam cracking of the lower alkanes, naphtha, or gas oil). About 1 million tons is available from this source. The hydrodealkylation of toluene is also a source of benzene depending on relative availability and demand. In 1978 a total of 1.37 million tons of benzene was produced from toluene. In the toluene hydrodealkylation process, toluene and toluene/xylene concentrates are reacted with hydrogen at about 730°C to give benzene and methane. Benzene is also obtained from toluene by transalkylation. This disproportionation reaction produces benzene and xylenes. The amount of benzene separated from coking of coal (about 8%) has decreased in recent years mainly because of the reduced production of steel in the United States.

---

* Noncontract prices fluctuate and affect derived chemicals accordingly.

3. Ethylbenzene

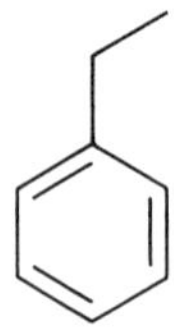

Consumption (1983):	United States	3,571,000 tons
	Western Europe	2,665,000 tons
	Japan	1,175,000 tons
Production Process:	Alkylation of benzene with ethylene	
Major Uses:	Manufacture of styrene monomer	
Price:	\$ 0.30/kg	

Ethylbenzene is manufactured almost exclusively by the alkylation of benzene with ethylene. A small amount is obtained in the United States by separation from the C_8 aromatic stream. The ethylation of benzene is conducted either in the liquid phase in the presence of a Friedel-Crafts catalyst ($AlCl_3$, BF_3, $FeCl_3$) or in the vapor phase with H_3PO_4 carrier catalysts or aluminum silicates. A process developed by Monsanto claims a 99% selectivity if the reaction is conducted at 140–200°C and 3–10 bar in the presence of $AlCl_3$.

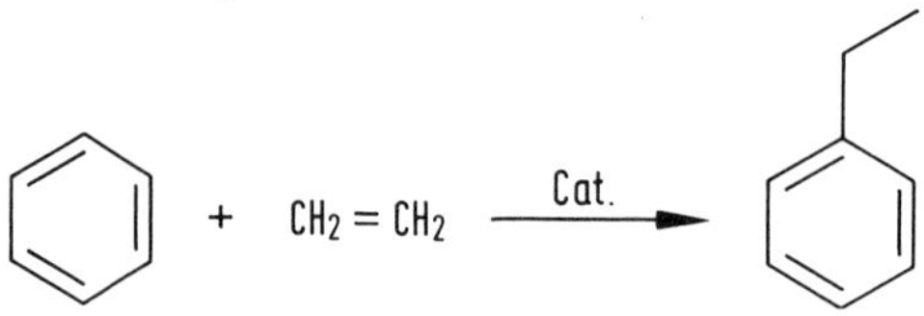

The vapor-phase reaction of benzene with ethylene is conducted at 300°C and 40–65 bar, using acidic carrier catalysts. All of the ethylbenzene produced is used in the manufacture of styrene monomer.

4. Styrene

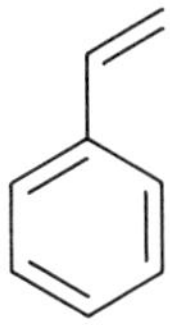

Consumption (1984):	United States	3,504,000 tons
	Western Europe	2,591,000 tons
	Japan	1,421,000 tons

Production Process: Dehydrogenation of ethylbenzene; coproduction with propylene oxide

Major Uses: Polystyrene and copolymers (80%), styrene/butadiene rubbers (10%)

Price: $ 0.53/kg

Styrene monomer is produced primarily by the catalytic dehydrogenation of ethylbenzene, which is conducted over an iron oxide catalyst containing Cr_2O_3 and KOH as promoters, at 550–600°C. After rapid cooling and addition of a polymerization inhibitor, styrene is separated from toluene and benzene by vacuum distillation.

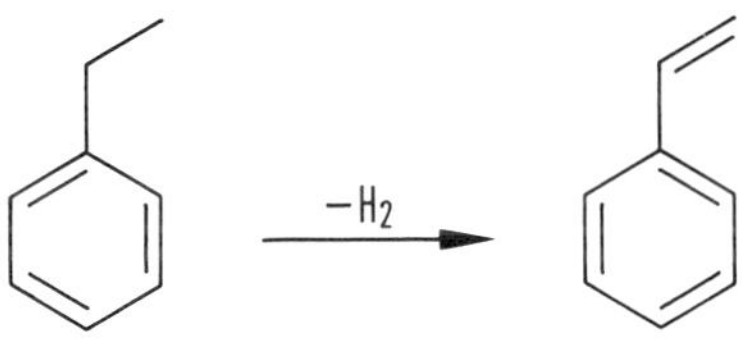

Dehydrogenation of ethylbenzene accounts for about 90% of the world styrene production. Air oxidation and reaction with propylene followed by dehydration of α-phenylethanol to styrene accounts for the remaining 10%.

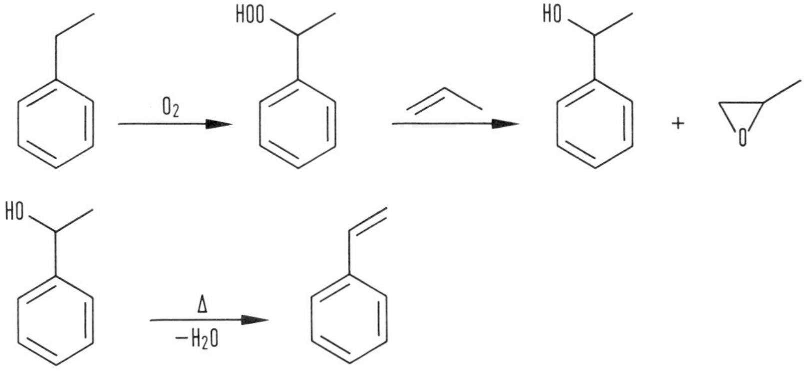

A small amount of styrene is also obtained by extraction of pyrolysis gasoline. Pyrolysis gasoline contains 6–8% styrene.

Styrene monomer is used to produce a range of homo-, co-, and terpolymers. The worldwide importance of styrene-based polymers is shown in Table 30.

Table 30. Consumption of Styrene-Based Polymers, 1984 (tons)

	PS	ABS[a]	SBR[b]
United States	1,735,000	476,000	890,000
Western Europe	1,661,000	350,000	654,000
Japan	768,000	368,000	302,000

[a] Acrylonitrile-butadiene-styrene terpolymer.
[b] Styrene-butadiene rubber.

5. Cumene

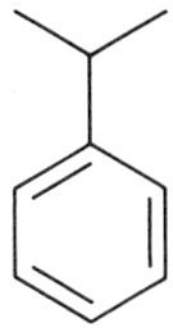

Consumption (1983):	United States	1,501,000 tons
	Western Europe	1,201,000 tons
	Japan	320,000 tons
Production Process:	Alkylation of benzene with propylene	
Major Uses:	Manufacture of phenol and acetone	
Price:	$ 0.50/kg	

Isopropylbenzene (cumene) is the reaction product of benzene and propylene, and it is used solely to produce phenol and acetone by the Hock process. The Hock process is a typical example of a brilliant compromise. The direct oxidation of benzene to phenol is economically unattractive because of low yield and low conversion. In contrast, the indirect air oxidation of isopropylbenzene proceeds in high selectivity, but for each pound of phenol another 0.6 lb of acetone is produced. Since acetone has major uses, the process is still in balance, and today almost all phenol is produced by the cumene process.

The propoxylation of benzene proceeds either in the liquid phase or in the vapor phase using Friedel-Crafts or acidic carrier catalysts (H_3PO_4/SiO_2). The liquid-phase reaction proceeds at 35–70°C and 7 bar, using $AlCl_3$ as the catalyst. The vapor-phase reaction proceeds at 250–300°C and 3–10 bar, using H_3PO_4/SiO_2 as the catalyst.

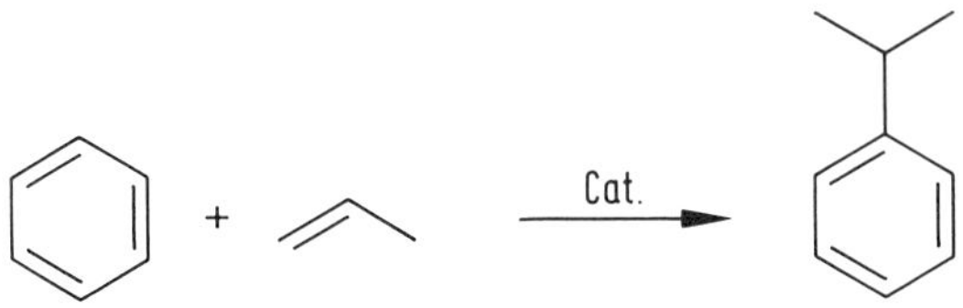

The propoxylation of benzene has also been used to manufacture hydroquinone and resorcinol from *p*-diisopropylbenzene and *m*-diisopropylbenzene, respectively.

Dehydrogenation of cumene affords α-methylstyrene, another building block for styrene plastics.

6. Phenol

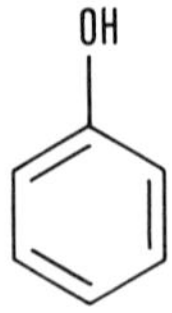

Consumption (1984):	United States	1,313,000 tons
	Western Europe	1,019,000 tons
	Japan	272,000 tons
Production Process:	Air oxidation of cumene (98%); oxidation of toluene	
Major Uses:	Phenolic resins (40%), bisphenol A (25%) caprolactam (15%), xylenols (7%), aniline (4%)	
Price:	\$ 0.80/kg	

Phenol is one of the more important secondary products derived from benzene. It is produced mainly by air oxidation of cumene, but it can also be obtained as a by-product from tar, coke oven water, and waste water from cracking units. A wide variety of synthetic processes have been employed, but today 98% of the phenol produced in the United States comes from the cumene process. This coproduct process was pioneered by Hock and Lang in 1944. The steps involved are air oxidation of cumene to its hydroperoxide and acid-catalyzed rearrangement of the hydroperoxide to phenol and acetone.

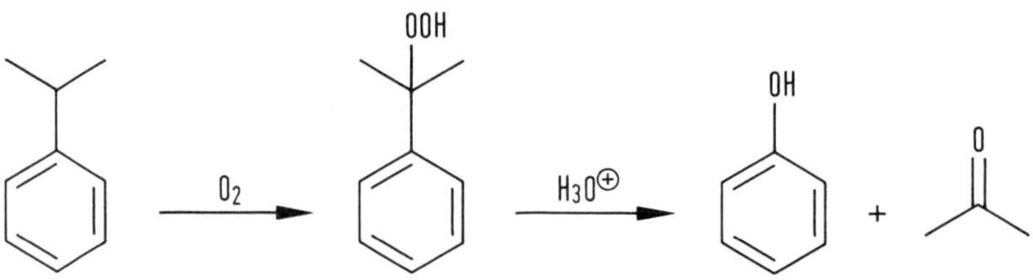

The conversion of cumene to the hydroperoxide is on the order of 35–40% to limit secondary reactions. Sodium carbonate is used as a stabilizer. The acid rearrangement is usually conducted with dilute sulfuric acid. By-products include α-methylstyrene and acetophenone.

The manufacture of phenol from toluene involves air oxidation to benzoic acid, followed by oxydecarboxylation (see p. 159).

Major uses of phenol include the phenolic resins. Approximately 40% ends up in housing. Phenolic adhesives are used for plywood, particleboard, waferboard, and fiberglass insulation. Phenolic foam is used in building insulation. Reaction of phenol with acetone affords bisphenol A, a monomer for polycarbonates, epoxy resins, and polysulfone. Caprolactam, the monomer for nylon-6, can also be obtained from phenol (see p. 142). Other polymer uses include polyphenylene oxide, which is produced by oxidative coupling of 2,6-dimethylphenol. The latter is obtained by the alkylation of phenol with methanol. Chemical uses of phenol include salicylic acid, pentachlorophenol, alkyl phenols, and phenol-derived insecticides. The alkylation of phenol with methanol produces a mixture of *o*- and *p*-cresol. *o*-Cresol is used in the manufacture of herbicides and insecticides.

Formation of 2,6-dimethylphenol, the monomer for polyphenylene oxide, is favored at higher temperatures and pressures using Al_2O_3 as the catalyst.

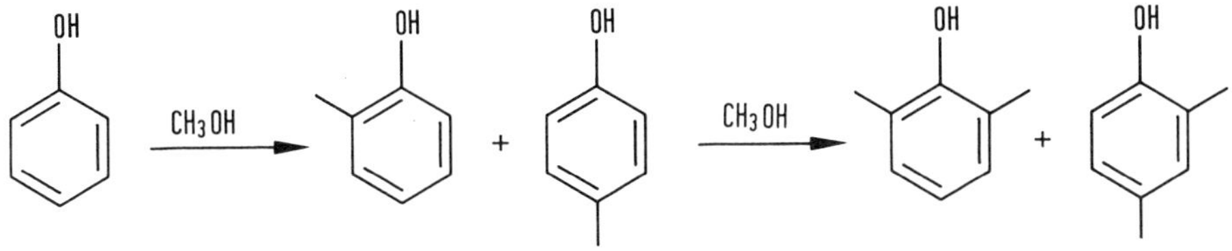

7. Cyclohexane

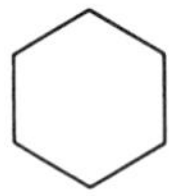

Consumption (1984):	United States	968,000 tons
	Western Europe	856,000 tons
	Japan	492,000 tons
Production Process:	Hydrogenation of benzene; isolation from oil refineries	
Major Uses:	Adipic acid (60%), caprolactam (30%)	
Price:	$ 0.30/liter	

Cyclohexane is produced mainly by catalytic hydrogenation of benzene. In the United States about 13% of the cyclohexane consumed domestically is isolated from gasoline production. The catalytic hydrogenation of benzene is conducted in the liquid phase (170–230°C and 20–40 bar) using nickel or platinum metal catalysts. Vapor-phase processes have also been developed. The direct hydrogenation/hydrodealkylation of toluene to cyclohexane is also used when economic conditions are favorable.

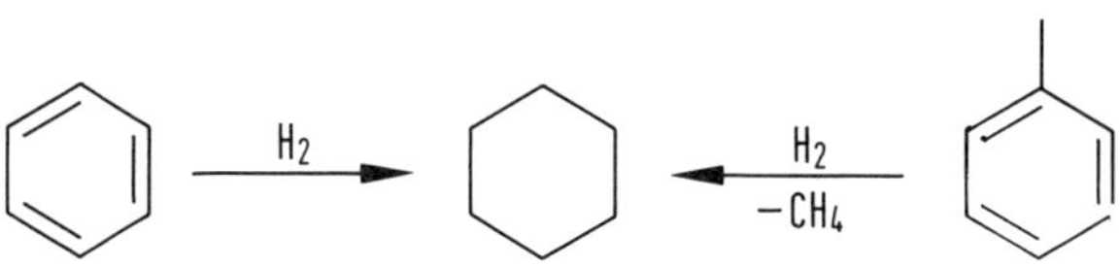

Most of the cyclohexane produced is air oxidized to a mixture of cyclohexanone and cyclohexanol, which ist the major feedstock for adipic acid and caprolactam. Reaction of cyclohexanol with ammonia over an alumina catalyst yields cyclohexylamine. Most of the cyclohexanone is used to produce caprolactam (see p. 142). Oxidation of the cyclohexanone/cyclohexanol mixture with nitric acid affords adipic acid (see p. 139). Nearly two-thirds of the cyclohexane produced in the United States is consumed in the manufacture of adipic acid, the monomer for nylon-6,6 and slightly less than one-third in the production of caprolactam. Cyclohexane competes with phenol as the basic raw material for caprolactam, the nylon-6 monomer.

8. Adipic Acid

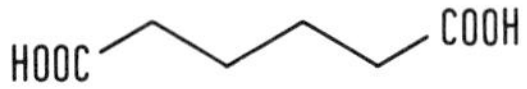

Consumption (1986):	United States	674,000 tons
	Western Europe	553,000 tons
	Japan	88,000 tons
Production Process:	Air oxidation of cyclohexane; hydrogenation of phenol followed by oxidation with nitric acid	
Major Uses:	Nylon-6,6 fibers (74%), nylon-6,6 resins (9%), hexamethylenediamine (8.5%), polyurethane resins (3.5%)	
Price:	\$ 1.15/kg	

Adipic acid is the key intermediate for the manufacture of nylon-6,6. It is derived from cyclohexan (the hydrogenation product of benzene) by air oxidation. The air oxidation of cyclohexane is conducted at 125–165°C and 8–15 bar in the presence of manganese or cobalt naphthenate. The cylohexane hydroperoxide intermediate is converted to a mixture of cyclohexanone and cyclohexanol (KA oil), using acid catalysis. The conversion is on the order of 10%, and the selectivity to ketone/alcohol is on the order of 80–85%.

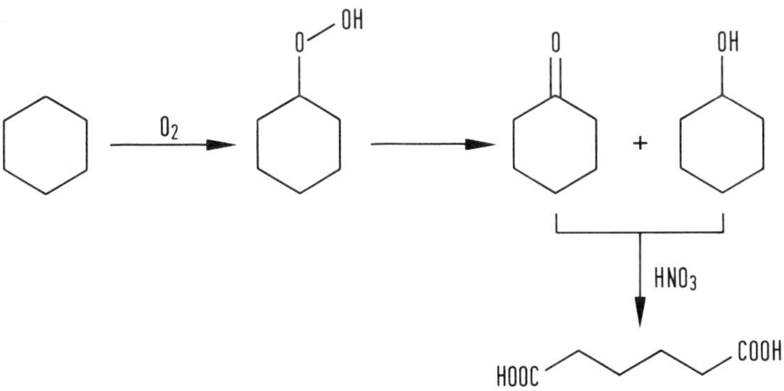

The oxidation of the KA oil is conducted with 60% HNO_3 at 50–80°C, but air oxidation is also feasible. The second approach to adipic acid involves the hydrogenation of phenol to a mixture of cyclohexanone and cyclohexanol, followed by nitric acid oxidation.

Adipic acid is used mainly as a monomer for nylon-6,6. It is also used as the precursor of hexamethylenediamine, the other nylon-6,6 monomer (see p. 140). Nonfiber uses of adipic acid include polyesterdiols (ethylene adipate, butylene adipate, etc.) for polyurethane applications, and adipic acid esters used as vinyl plasticizers.

9. Hexamethylenediamine 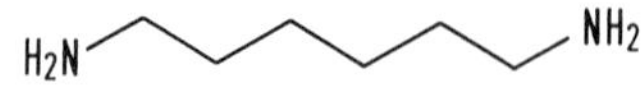

Consumption (1982):	United States	408,000 tons
	Western Europe	230,000 tons
	Japan	30,000 tons
Production Process:	Hydrogenation of adiponitrile	
Major Uses:	Nylon-6,6 (99%), polyurethanes	
Price:	$ 2.55/kg	

Only Monsanto uses adipic acid for the manufacture of hexamethylenediamine in the United States. The ammonium salt of adipic acid is dehydrated to adiponitrile. This reaction is conducted in the melt at 200–300°C in the presence of phosphoric acid as the catalyst. Hydrogenation of adiponitrile yields hexamethylenediamine.

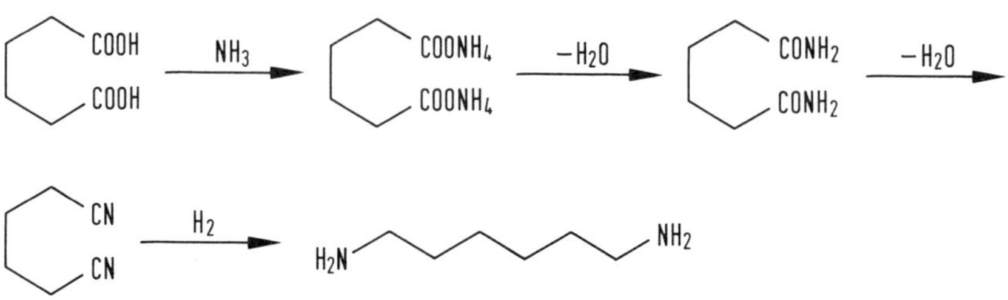

A second process based on a mixture of adipic acid and 5-hydroxycaproic acid was developed by Celanese. Reduction of this mixture affords 1,6-hexanediol. Esterification of the acid mixture with 1,6-hexanediol and hydrogenation of the esters converts the total mixture into 1,6-hexanediol. Hexanediol is treated with ammonia over an alumina catalyst at 210°C and 225 bar to give hexamethylenediamine. In 1981 Celanese abandoned this approach by closing its plant in Bay City, Texas.

Dupont has based its hexamethylenediamine manufacture on butadiene as feedstock. Formerly, oxychlorination of butadiene was followed by the reaction of 1,4-dichlorobutene with sodium cyanide to give 1,4-dicyano-2-butene, which was hydrogenated to the final product. This technology was recently replaced by the direct reaction of butadiene with hydrogen cyanide, eliminating the use of chlorine and sodium hydroxide.

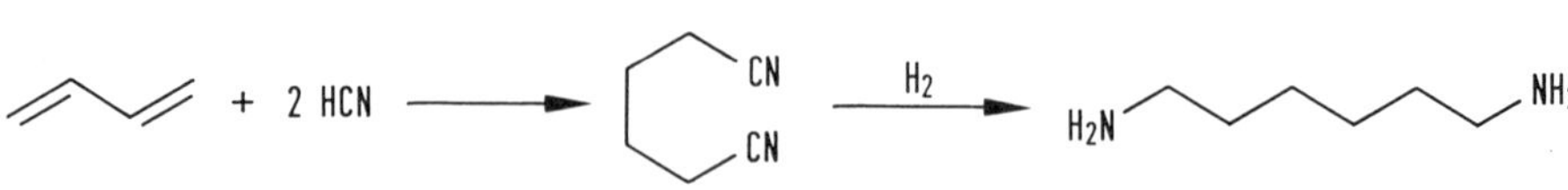

Monsanto developed a route to adiponitrile based on propylene. In this approach acrylonitrile was subjected to an electrohydrodimerization reaction to give adiponitrile. However, this route was abandoned because of the high cost of electricity.

The hydrogenation of adiponitrile, which is common to all commercial processes with the exception of the recently abandoned Celanese process, is conducted at 100−135°C and 600−650 bar using a cobalt/copper catalyst or at 100−180°C and 300−350 bar using an iron catalyst. The formation of polyamines and hexamethyleneimine is repressed by adding ammonia. The selectivity to hexamethylenediamine is on the order of 90−95%.

Hexamethylenediamine is also used as a monomer for the manufacture of specialty nylons, such as nylon-6,10 (hexamethylenediamine and sebacic acid), nylon-6,9 (hexamethylenediamine and azelaic acid), and nylon-6,12 (hexamethylenediamine and 1,12-dodecanedioic acid). Another minor use of hexamethylenediamine is in the formulation of light-stable polyurethane coatings. Phosgenation of hexamethylenediamine affords hexamethylene diisocyanate, which is used in form of its derivatives because of the low vapor pressure of the diisocyanate. A trifunctional biuret derivative is obtained from hexamethylene diisocyanate and water. A second approach involves trimerization of hexamethylene diisocyanate.

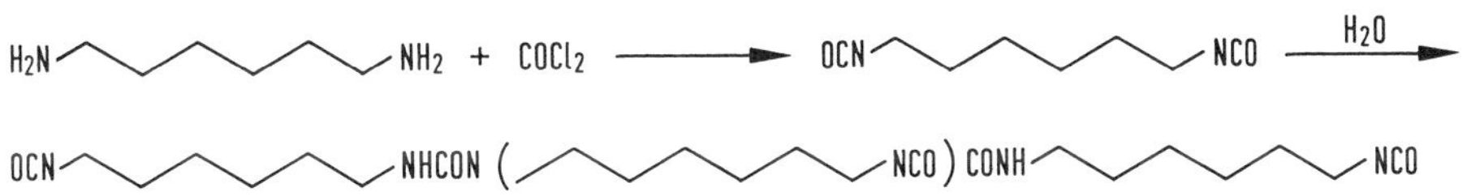

The trifunctional biuret and the higher functional trimers (isocyanurates) are used in rigid industrial coating applications where color stability and abrasion resistance are required.

10. Caprolactam

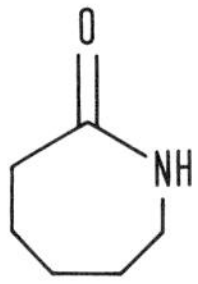

Consumption (1983):	United States	445,000 tons
	Western Europe	478,000 tons
	Japan	436,000 tons
Production Process:	Beckmann rearrangement of cyclohexanone oxime; cyclohexane and nitrosyl chloride; hexahydrobenzoic acid and nitrosylsulfuric acid.	
Major Uses:	Nylon-6 fiber (93%), nylon-6 resins (7%)	
Price:	$ 1.70/kg	

Approximately 73% of caprolactam capacity is based on cyclohexane feedstock, 23% is based on phenol, and 4% is based on toluene. Cyclohexanone, the key intermediate for caprolactam, is produced either by air oxidation of cyclohexane or by hydrogenation of phenol. Reaction of cyclohexanone with hydroxylamine sulfate followed by treatment with oleum produces caprolactam. The by-product is ammonium sulfate.

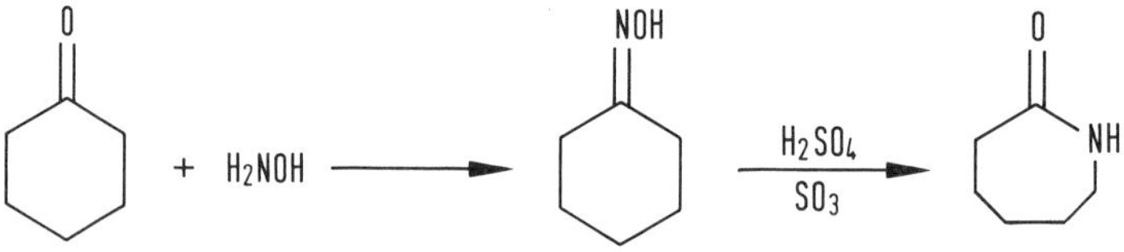

For each pound of caprolactam, 4.4 lb of ammonium sulfate is produced. The driving force for the development of new technology was the reduction or elimination of this by-product. For example, BASF produces hydroxylamine by reducing nitrogen oxide (NO) with hydrogen, thereby eliminating the ammonium sulfate by-product. The Stamicarbon HPO process also avoids ammonium sulfate because hydroxylamine is generated and reacted in situ. In Japan, Toray has developed a different route to cyclohexanone oxime involving the photochemical nitrosation of cyclohexane with nitrosyl chloride.

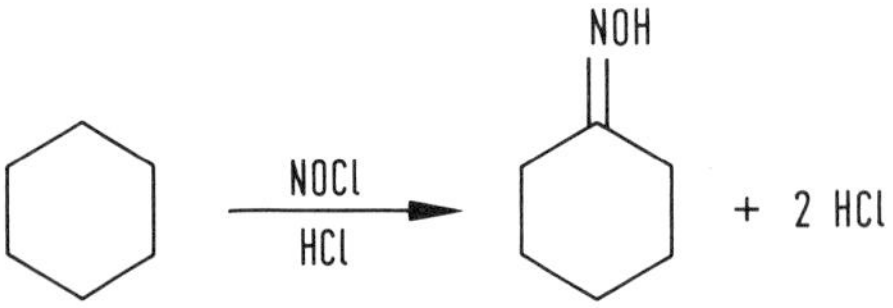

In Italy, Snia-Viscosa has developed a new process based on toluene. The first step, air oxidation of toluene to benzoic acid, is well established. Reduction of benzoic acid to hexahydrobenzoic acid followed by treatment with nitrosylsulfuric acid gives caprolactam and carbon dioxide.

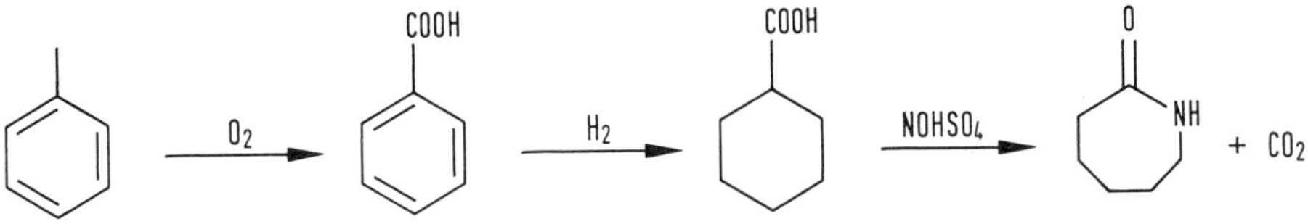

UCC developed a process based on the air oxidation of acetaldehyde to peracetic acid, which was used to oxidize cyclohexanone to caprolactone. Treatment of caprolactone with ammonia produces caprolactam. This process never became commercial, but caprolactone is still manufactured in this manner. Caprolactone is a building block for thermoplastic polyurethane elastomers.

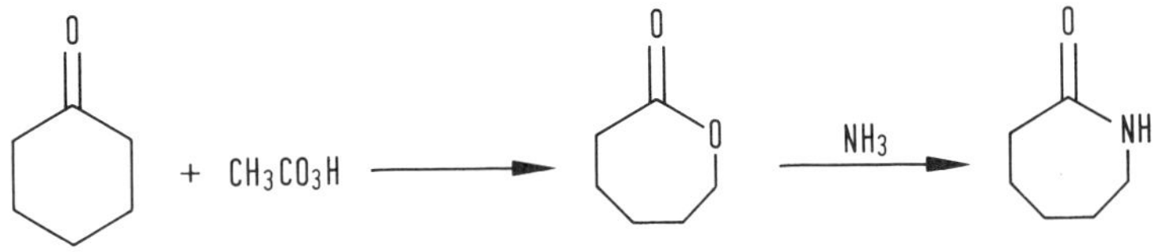

11. Aniline

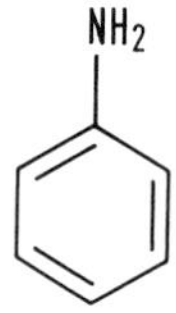

Consumption (1983):	United States	302,000 tons
	Western Europe	280,000 tons
	Japan	78,000 tons
Production Process:	Reduction of nitrobenzene; amination of phenol	
Major Uses:	Polymeric isocyanates (59%), rubber chemicals (23%), dyes and pigments (5%), hydroquinone (3%)	
Price:	$ 0.90/kg	

The classical method of synthesis of aniline is the reduction of nitrobenzene with hydrogen in the vapor phase using a copper/silica catalyst. Nitrobenzene is produced in the nitration of benzene using a mixture of nitric acid and sulfuric acid.

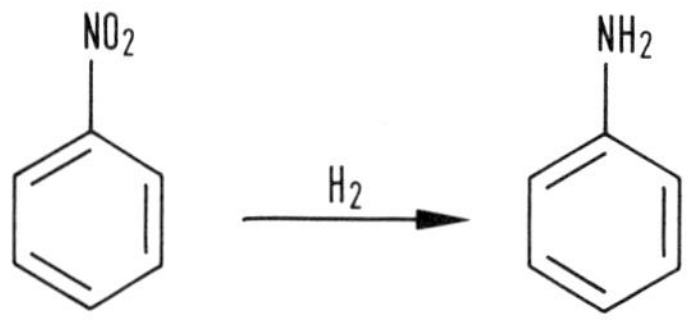

Aniline is also obtained by Mobay as a by-product in the production of iron oxide pigment. Recently Aristech Chemicals completed a 90,000-ton aniline plant using phenol as the feedstock. This technology was developed by Halcon, and it is also used by Mitsui Toatsu in Japan. The amination of phenol is conducted in the vapor phase using an alumina catalyst, and yields are very high (99%).

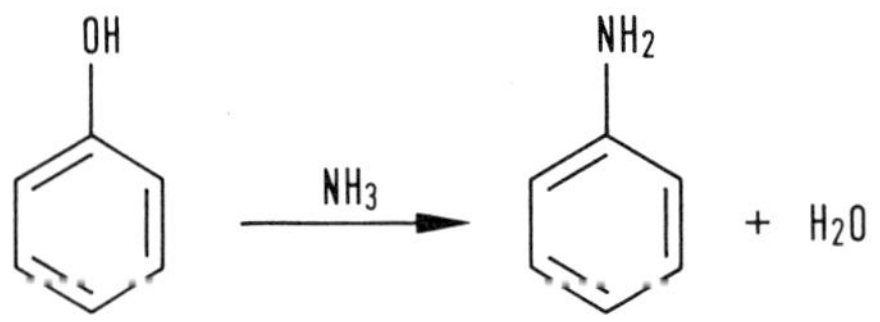

The major use of aniline is in the manufacture of polyamines via condensation with formaldehyde. This reaction is conducted in the liquid phase with hydrochloric acid as the catalyst. The reaction product is a mixture of oligomeric amines, containing about 50% 4,4'-diaminodiphenylmethane (MDA). The polyamines are subsequently phosgenated to give a mixture of oligomeric isocyanates, from which the diisocyanate (MDI) can be separated by distillation.

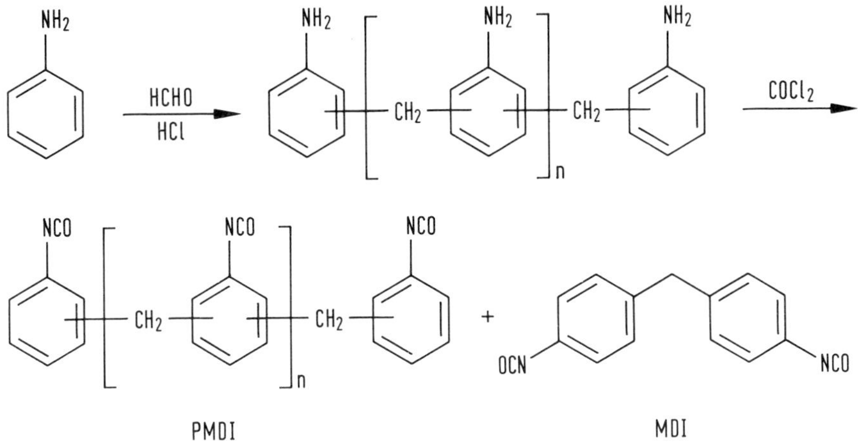

The diisocyanate (MDI) is used in the production of polyurethane elastomers and in the formulation of coatings, sealants, and adhesives. Elastomers have found large-scale applications in the United States automotive industry for the manufacture of exterior parts such as bumpers, fenders, and body panels. In 1986 about 40,000 tons of RIM (reaction injection molded) polyurethane parts based on MDI were produced in the United States. The polymeric isocyanate is used in the manufacture of rigid polyurethane foam, which is widely used as insulation material in the building and construction industry. Rigid polyurethane foams are blown with fluorocarbons to produce a closed-cell structure with excellent insulation properties. The amounts of rigid polyurethane foam produced in major world markets are listed in Table 31.

Table 31. Consumption of Rigid Polyurethane Foam, 1984 (tons)

United States	309,000
Western Europe	250,000
Japan	93,000

The comonomers for the manufacture of rigid polyurethane foams are polyether and polyester polyols. The polyurethanes are produced from liquid monomers at room temperature or slightly elevated temperature, which facilitates the processing.

Aniline is also used as a raw material for rubber chemicals (antioxidants, accelerators, activators, and vulcanizers). Benzthiazoles are the more important accelerators, and *p*-phenylenediamine derivatives are used as antioxidants. For example, benzthiazol manufactured from aniline and carbon disulfide is oxidized to its disulfide (MBTS), the most important benzthiazol accelerator.

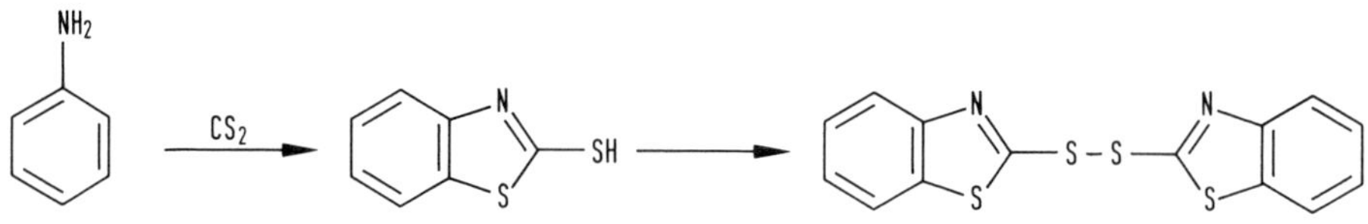

Diphenylguanidine, used as an activator for the benzthiazole accelerators, is produced from cyanogen chloride and aniline.

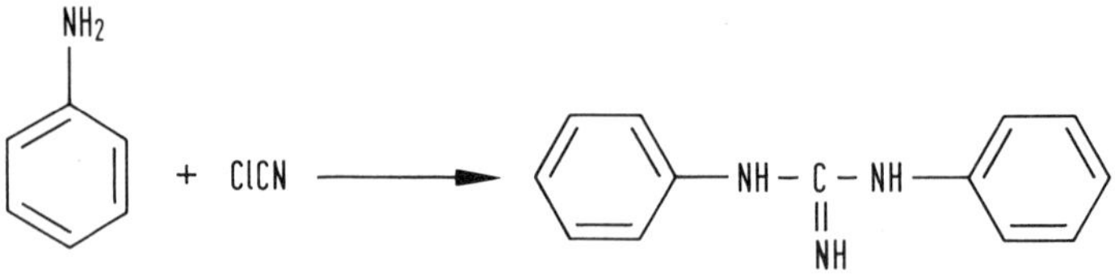

A number of important dyes and some pigments are based on aniline. Oxidation of aniline with manganese dioxide gives benzoquinone, which is reduced to hydroquinone, an important rubber chemical and photographic developer.

Other aromatic amines used in the production of industrial polymers are shown in Table 32. *m*-Phenylenediamine is obtained by the reduction of *m*-dinitrobenzene. *p*-Phenylenediamine is produced by the hydrogenation of *p*-nitroaniline. Oxydianiline is manufactured from *p*-nitrophenol and *p*-nitrochlorobenzene, followed by reduction.

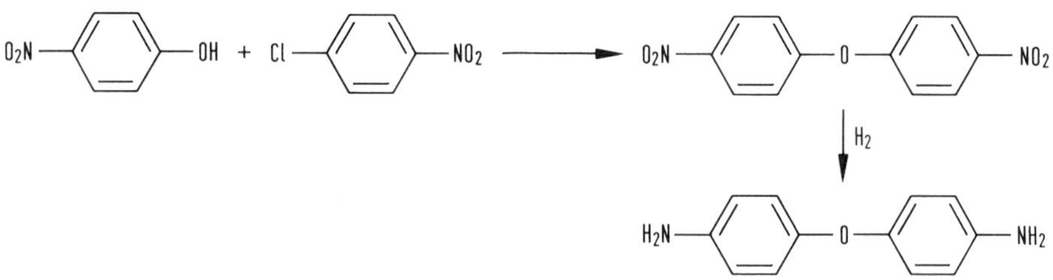

Tetraaminobiphenyl is produced from dichlorobenzidine and ammonia.

Table 32. Aromatic Amine Monomers Used in Industrial Polymers

Monomer		Derived Polymers
p-Phenylenediamine		Aramids (Kevlar)
m-Phenylenediamine	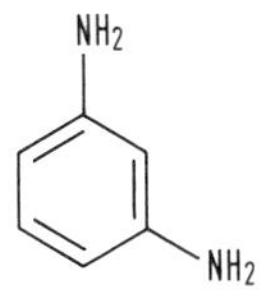	Aramids (Nomex), Polyimide (Ultem)
4,6-Dihydroxy *m*-phenylenediamine	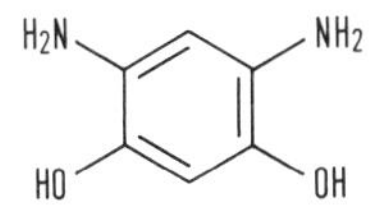	Polybenzoxazole
Bisamine A	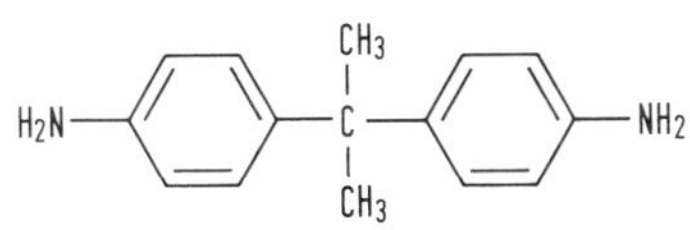	Transparent polyamides
4,4'-Diaminodicyclo-hexylenemethane	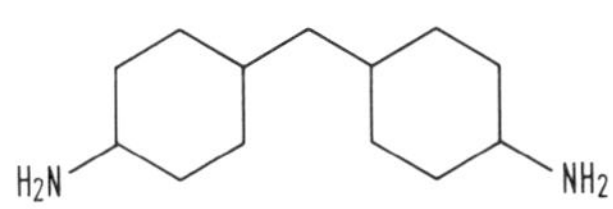	Transparent polyamides
4,4'-Diamino-3,3'-dimethyl-dicyclohexylenemethane	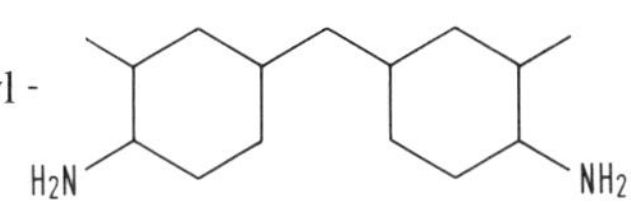	Transparent polyamides
Oxydianiline	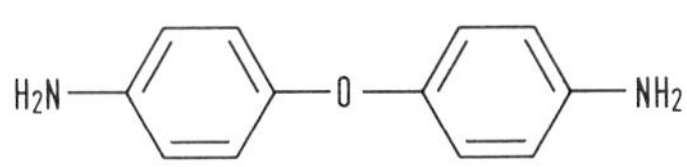	Polyimide (Kapton, Vespel)
Tetraaminobiphenyl	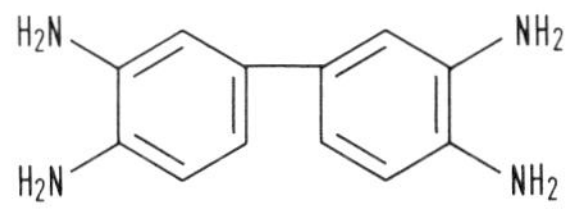	Polybenzimidazole (PBI)

12. Chlorobenzene

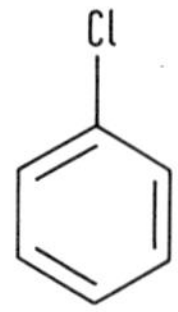

Consumption (1985):	United States	103,000 tons
	Western Europe	95,000 tons
	Japan	28,000 tons
Production Process:	Chlorination or oxychlorination of benzene	
Major Uses:	Solvent (26%), nitrochlorbenzenes (41%), diphenyloxide and *p*-phenylphenol (20%)	
Price:	\$ 0.92/kg	

The chlorination of benzene in the presence of Friedel-Crafts catalysts gives mixtures of monochlorobenzene (MCB), *o*-dichlorobenzene (ODCB), and *p*-dichlorobenzene (PDCB). The reaction is conducted at 25–50°C using $FeCl_3$ as the catalyst.

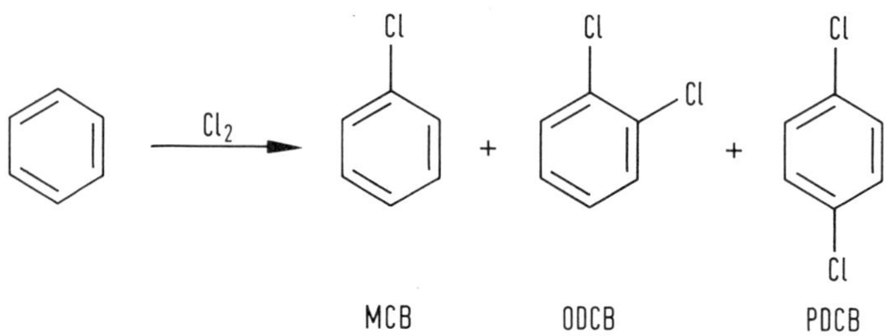

Chlorobenzenes are also obtained by oxychlorination of benzene with hydrogen chloride and air at approximately 250°C in the presence of $CuCl_2/FeCl_3/Al_2O_3$. The conversion is only on the order of 10–15% to repress the formation of dichlorobenzenes. The advantage of the oxychlorination process is the use of the abundantly available HCl, but the reaction is highly corrosive.

Monochlorobenzene is used mainly as a solvent, but other chemical derivatives such as nitrochlorobenzenes, diphenyloxide, *p*-phenylphenol, and DDT also use significant amounts of MCB. DDT is manufactured by the acid-catalyzed condensation of chlorobenzene with chloral.

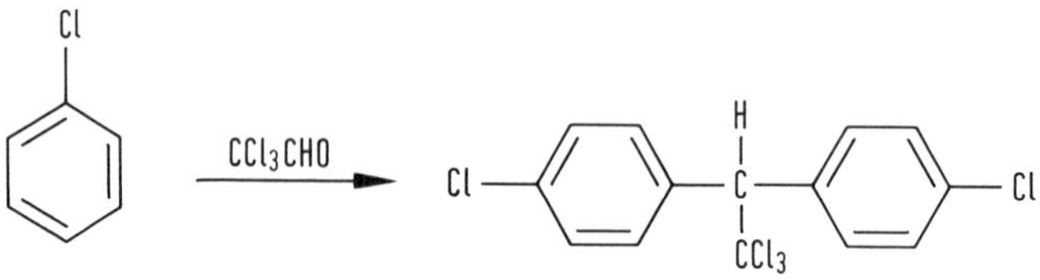

In the United States in 1978 about 11,000 tons of chlorobenzene was used in the manufacture of DDT solely for export because the use of DDT as an insecticide has been banned in the United States. Current world production of DDT is still on the order of 100,000 tons, because no effective substitute has been found to combat the anopheles mosquito, the carrier of malaria.

p-Dichlorobenzene is the basic building block for poly(phenylene sulfide). Reaction of *p*-dichlorobenzene with sodium sulfide in a polar solvent produces PPS, an important engineering plastic.

In the United States, approximately 10,000 tons of *p*-dichlorobenzene was used in the production of PPS in 1985.

13. Hydroquinone

Consumption (1984):	United States	14,000 tons
	Western Europe	–*
	Japan	4,000 tons
Production Process:	Oxydation of aniline; air oxidation of diisopropylbenzene; oxidation of phenol	
Major Uses:	Rubber chemicals (35%), photographic developer (25%)	
Price:	$ 3.75/kg	

The classical route to hydroquinone is the oxidation of aniline with manganese dioxide to benzoquinone, followed by reduction with sulfur dioxide to give hydroquinone.

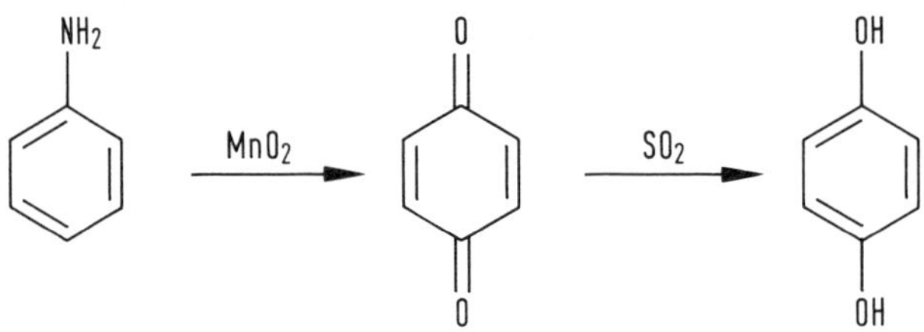

In view of the huge amounts of manganese salts produced in this process, it is not surprising that new processes for hydroquinone were developed.
Mitsui Petrochemicals in Japan and Goodyear in the United States now use a variation of the Hock process. Air oxidation of *p*-diisopropylbenzene affords the bishydroperoxide. which is rearranged in acid to give hydroquinone and the co-product acetone.

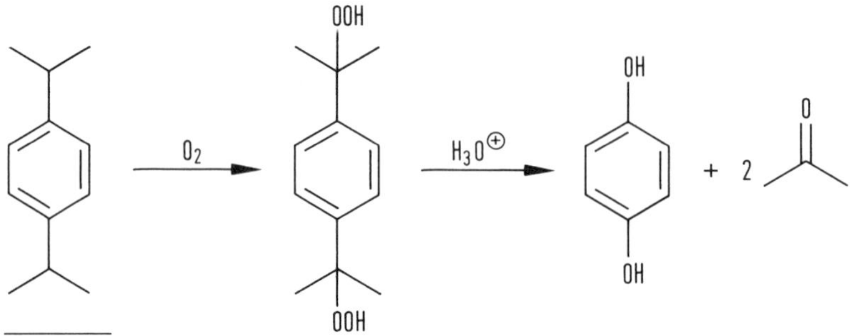

* not available.

Direct hydroxylation of phenol with hydrogen peroxide in the presence of formic acid is practiced by Rhône-Poulenc in France and Ube in Japan. This process produces a mixture of hydroquinone and catechol. Because of the smaller market for catechol, this process is of limited value.

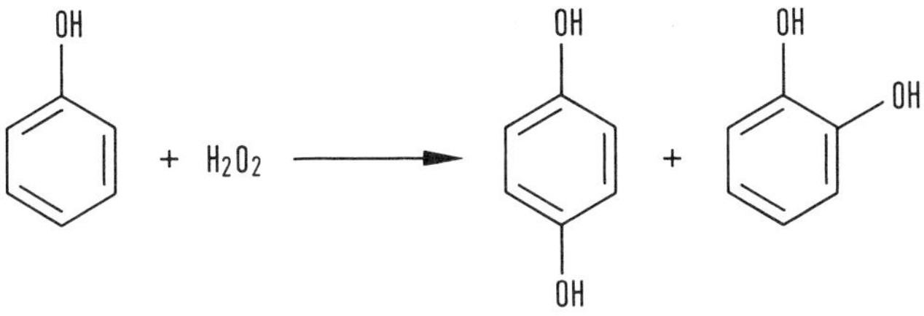

An emission-free hydroquinone process based on bisphenol A was developed by Upjohn. Pyrolysis of bisphenol A produces isopropenylphenol, which is oxidized with hydrogen peroxide at room temperature to give hydroquinone and acetone. Since the phenol and acetone that are generated are recycled, the overall reaction, amounts to a selective hydroxylation of phenol in the *p*-position.

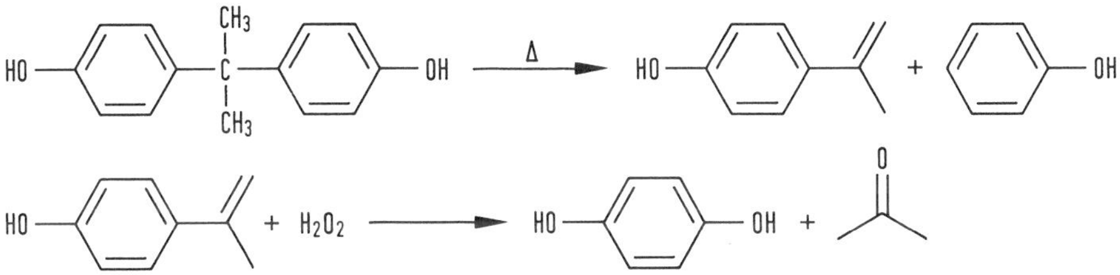

Hydroquinone is used as a photographic developer, a rubber antioxidant, and a polymerization inhibitor. Recently, hydroquinone was used by ICI as a monomer for their high-performance polymer PEEK. The monomethyl ether of hydroquinone is alkylated to *tert*-butylhydroxyanisole, a food antioxidant.

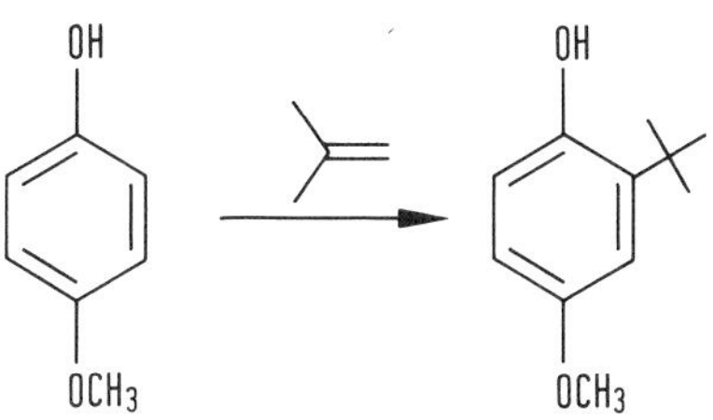

14. Resorcinol

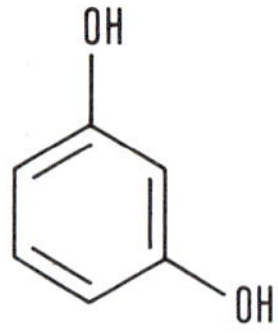

Consumption (1980):	United States	8 500 tons
	Western Europe	7 000 tons
	Japan	2 500 tons
Production Process:	Sulfonation of benzene followed by sodium hydroxide fusion; air oxidation of *m*-diisopropylbenzene	
Major Uses:	Rubber products (32%), wood products (16%), phenolic resins	
Price:	$ 3.45/kg	

Resorcinol is produced in the United States by the classical alkaline fusion process. Benzene is first sulfonated with either H_2SO_4/SO_3 or SO_3 to give benzene *m*-disulfonic acid, which is neutralized with sodium hydroxide. The sodium hydroxide fusion of the sodium salt of benzene *m*-disulfonic acid is conducted at 300°C.

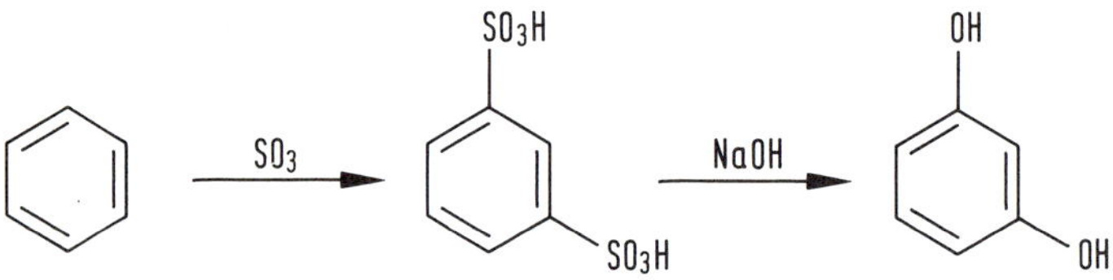

In view of the generation of large amounts of sodium sulfite and sodium sulfate, the classical process will eventually be replaced by an air-oxidation process similar to the Hock phenol synthesis. Mitsui in Japan is currently producing resorcinol by air oxidation of *m*-diisopropylbenzene.

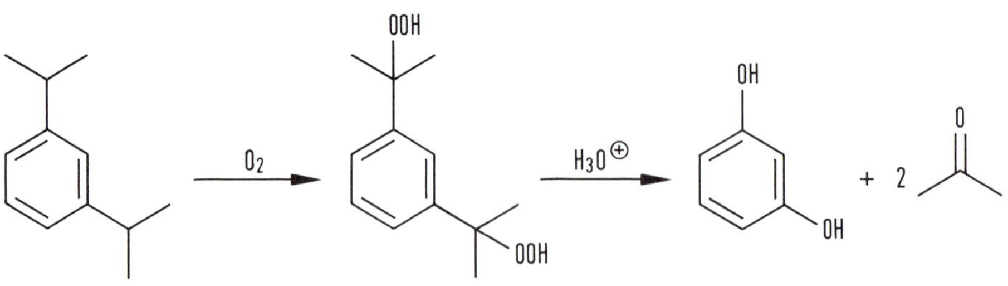

Resorcinol/formaldehyde copolymers are used primarily as tire cord adhesives. They are also used in the manufacture of wood laminates, especially when room-temperature curing is required. The comonomer used in the manufacture of resorcinol/formaldehyde resins is the less expensive phenol.

15. Bisphenol A

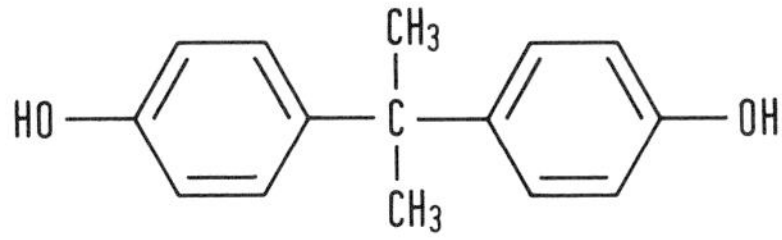

Consumption (1985):	United States	341,000 tons
	Western Europe	171,000 tons
	Japan	61,000 tons
Production Process:	Reaction of phenol with acetone	
Major Uses:	Polycarbonate (48%), epoxy resins (45%)	
Price:	\$ 0.95/kg	

Bisphenol A is produced by reacting phenol with acetone in the presence of an acid catalyst. Common catalysts are aqueous acids or acid clays. Promoters such as thioglycolic acid and resorcinol are also used.

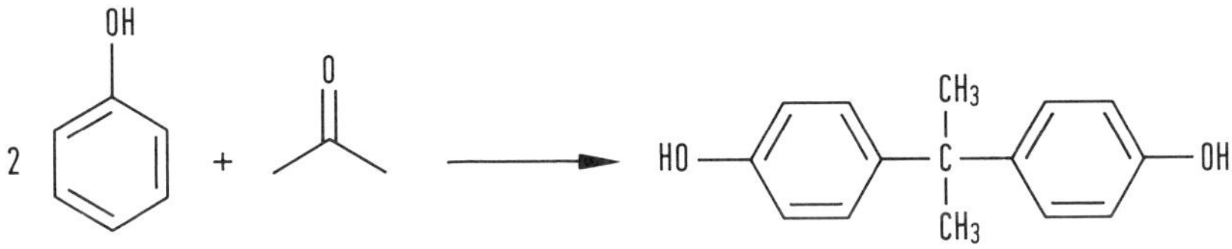

Bisphenol A is used mainly as a monomer for polycarbonate and for epoxy resins. Table 33 gives the consumption of polycarbonates and epoxy resins in the three major world markets.

Table 33. Consumption of Polymers Derived from Bisphenol A, 1984 (tons)

	Polycarbonate	Epoxy Resins
United States	180,000	186,000
Western Europe	65,000	100,000
Japan	31,500	60,000

Polycarbonates are produced from bisphenol A and phosgene by an interfacial polymerization process. Methylene chloride is used as the water-immiscible phase, and sodium hydroxide as the hydrogen chloride scavenger.

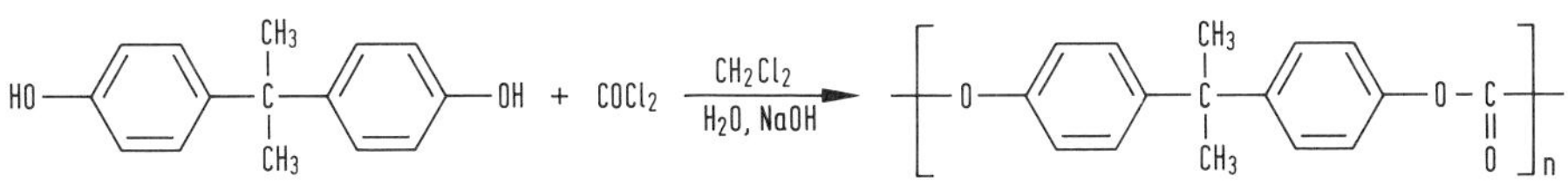

Epoxy resins are manufactured from bisphenol A and epichlorohydrin.

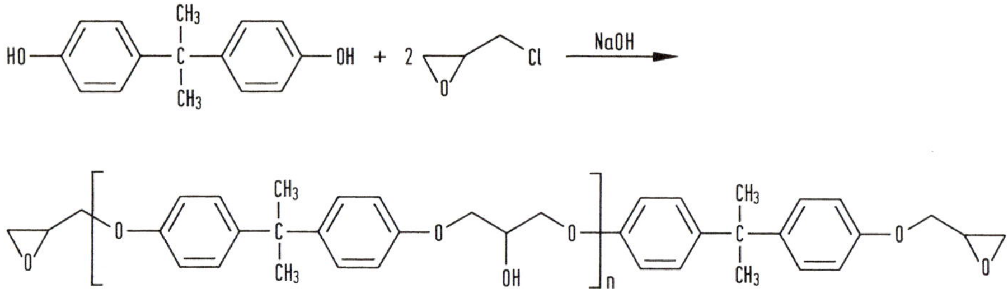

Another monomer use for bisphenol A is in the manufacture of polysulfone. In this process the sodium salt of bisphenol A is reacted with 4,4'-dichlorodiphenyl sulfone in DMSO to produce polysulfone.

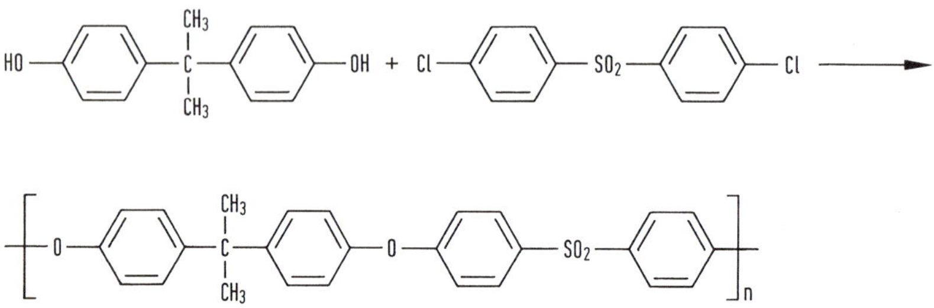

Polyester resins have been produced from bisphenol A, propylene oxide, and fumaric acid. Bisphenol A is also a building block for injection-moldable polyimides. Tetrabromobisphenol A is used as a fire retardant in polymers derived from bisphenol A.

Chapter VIII. Toluene Derivatives

1. Introduction

Although most of the important monomers for industrial polymers are based on benzene as the basic feedstock, research is under way to utilize the less expensive toluene whenever possible. An example is the manufacture of *p*-methylstyrene from ethylene and toluene. The derived *p*-methylstyrene has some advantages over polystyrene, but it remains to be seen if this technology will develop as rapidly as linear low-density polyethylene. Considerable research efforts are also in progress to manufacture basic chemicals such as phenol, styrene, and terephthalic acid from toluene.

A major established use for toluene is in the manufacture of TDI (tolylene diisocyanate), which is used in the manufacture of flexible polyurethane foam. Flexible polyurethane foam is widely used in furniture and bedding, as automotive seating, and as carpet underlay. The world consumption of flexible polyurethane foam in 1980 was slighthy over 1.8 million tons.

2. Toluene

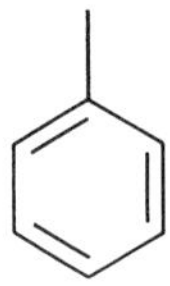

Consumption (1984):	United States	2,535,000 tons
	Western Europe	1,078,000 tons
	Japan	784,000 tons
Production Process:	Isolated from catalytic reformate; by-product in olefin and styrene production	
Major Uses:	Benzene (67%), solvent (17%), tolylene diisocyanate (7%)	
Price:	$ 0.29/kg	

A large amount of the toluene produced in petroleum refining operations is consumed in gasoline production. Most of the isolated toluene is obtained in the catalytic reformation of the naphtha fraction of crude oil. However, 1–2% of toluene is still obtained from coal tar. The largest single use of toluene is its hydrodealkylation to benzene, but this process depends on a sufficient price differential between toluene and benzene to be economically attractive. Disproportionation of toluene to benzene and xylenes is also conducted if a higher demand for these chemicals is indicated. With the reduction of the lead content in gasoline, toluene is used increasingly as an octane enhancer in gasoline.

The solvent use of toluene is mainly in the formulation of paints and coatings. Recently adopted United States standards that restrict the use of hydrocarbons in these applications have led to a decrease in the use of toluene as solvent. Toluene diisocyanate is produced by dinitration of toluene, reduction to toluenediamine, and phosgenation to the diisocyanate. Toluene diisocyanate (TDI) is used in the manufacture of open-cell flexible polyurethane foam.

Other chemical uses of toluene include the production of benzoic acid, benzyl chloride, benzaldehyde, and trinitrotoluene (TNT).

In view of the economic advantage of using toluene as the feedstock for major organic industrial chemicals, research is under way to substitute toluene for benzene. A recent example is the manufacture of *p*-methylpolystyrene from *p*-methylstyrene. The chemistry is very similar to the current route to styrene; toluene is reacted with ethylene to give *p*-methylethylbenzene, which is dehydrogenated to *p*-methylstyrene. The key to obtaining mainly the *p*-isomer is a ZSM-5 zeolite catalyst.

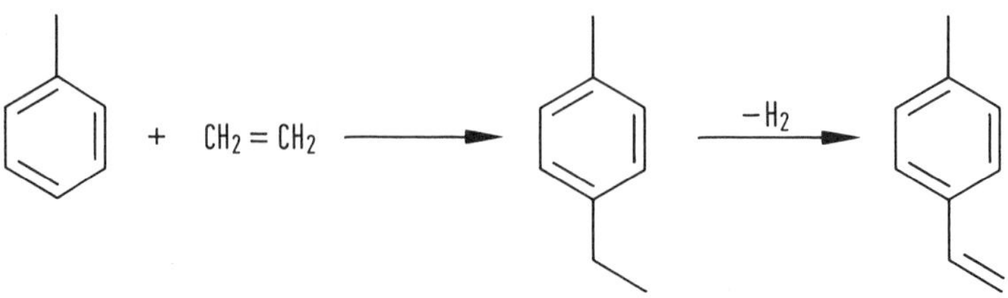

The derived *p*-methylpolystyrene has significant advantages over polystyrene, such as a higher glass transition temperature and better flow properties. A 16,000 ton plant for PMS is under construction by Mobil. A total of 45,000 tons/year of a mixture of isomeric vinyl toluenes is supplied by Dow and Cosden (*meta*/*para* ratio 2:1) for the manufacture of unsaturated polyesters, alkyd coatings, and adhesives. Attempts are also being made to produce styrene monomer from toluene. For example, reaction of toluene with syngas (via benzyl alcohol) or methanol produces styrene. Also, oxidative coupling of toluene to stilbene, followed by disproportionation, yields styrene (Scheme 3). All processes are still in the pilot-plant stage.

Scheme 3

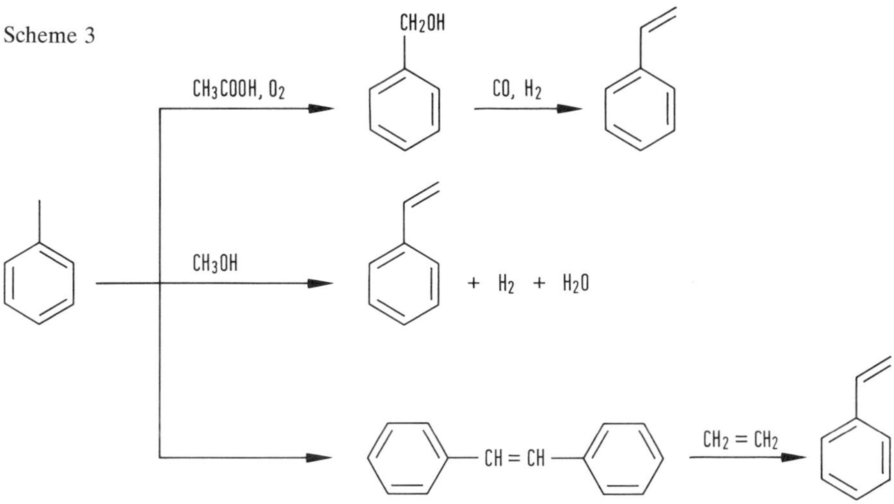

The stilbene route developed by Monsanto involves a novel disproportionation. However, the selectivity to stilbene is low, and it is difficult to handle stilbene because of its high melting point. The methylation of toluene with methanol in the presence of a zeolite catalyst is also under investigation by Monsanto. In the reaction of toluene with syngas, the first step is conversion of toluene to benzyl alcohol, followed by further reaction with syngas to give α-phenethyl alcohol. The latter is dehydrated to the styrene monomer.

The manufacture of phenol from toluene is practiced by one company (Kalama) in the United States. In this process, toluene is initially oxidized to benzoic acid, which is oxidatively decarboxylated to phenol.

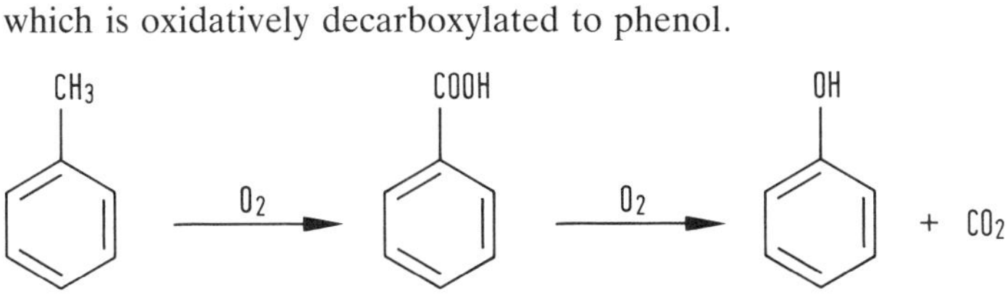

The first step involves air oxidation of toluene to benzoic acid. This reaction can be conducted in the liquid phase with air at 110–120°C and 2–3 bar in the presence of a cobalt catalyst. The oxydecarboxylation of benzoic acid is conducted in the melt (220–250°C) in the presence of steam and air using a copper salt as the catalyst. The intermediate in this reaction is copper benzoate.

Palladium-catalyzed acetoxylation of toluene to produce cresols has been attempted.

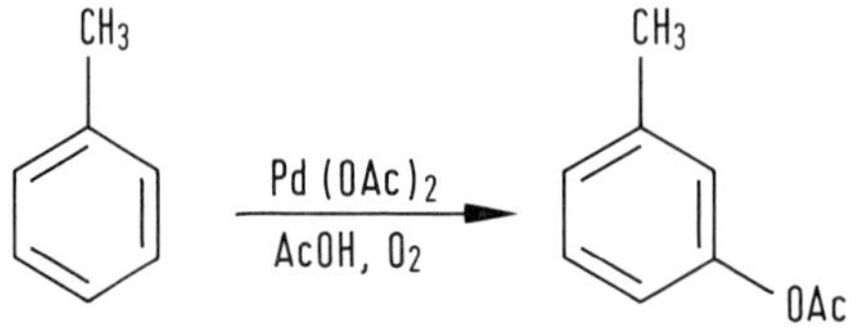

A toluene route to terephthalic acid has been piloted by Mitsubishi. In this approach, toluene is reacted with carbon monoxide in the presence of HF/BF_3 to give *p*-tolylaldehyde, which upon oxidation produces terephthalic acid.

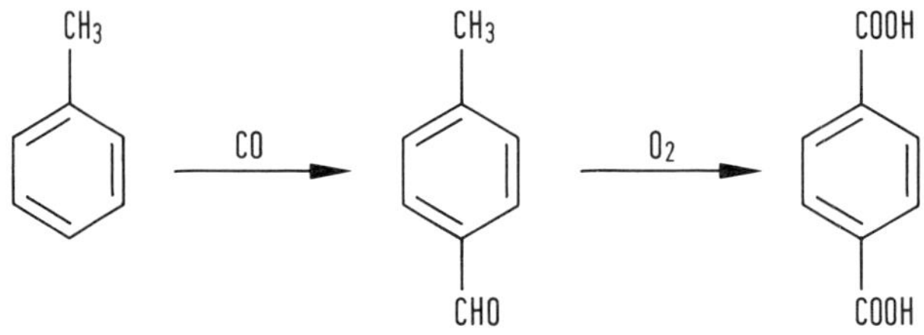

3. Benzyl Chloride

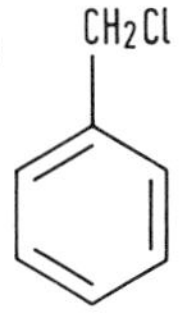

Consumption (1982):	United States	39,000 tons
	Western Europe	40,000 tons
	Japan	8,000 tons
Production Process:	Chlorination of toluene	
Major Uses:	Benzyl phthalates (61%), benzyl alcohol (11%), quaternary ammonium salts (10%)	
Price:	\$ 1.30/kg	

The direct chlorination of toluene is the method of choice for the manufacture of benzyl chloride. The yield is about 90%, and benzal chloride and benzotrichloride are byproducts.

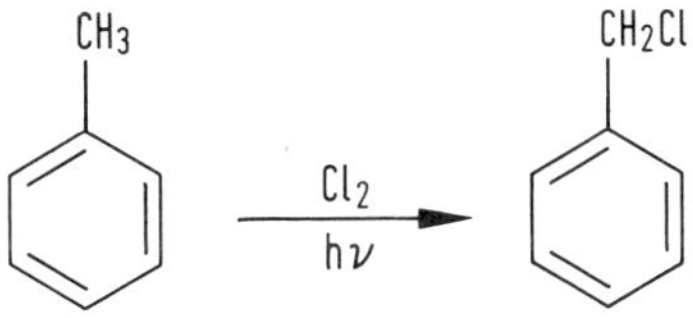

Benzyl chloride is used mainly to produce benzyl phthalate plasticizers. The major product is *n*-butyl benzyl phthalate, which is used in vinyl flooring. Benzyl alcohol, the second largest outlet for benzyl chloride, is made by hydrolyzing benzyl chloride with sodium carbonate. Benzyl alcohol is used in developing baths for color film and color transparencies; as a dispersing agent for silver halides in mixed packet emulsions; for perfumes, soaps, and cosmetics; and as a solvent for pharmaceuticals. Benzyl quaternary ammonium salts are used as germicides and disinfectants.

4. Benzoic Acid

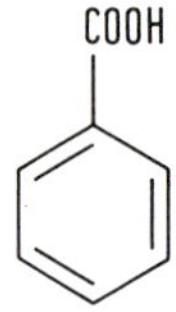

Consumption (1980):	United States 80,000 tons
Production Process:	Oxidation of toluene; decarboxylation of phthalic anhydride
Major Uses:	Phenol (50%), plasticizers (18%), benzoyl chloride (12%), sodium benzoate (19%), caprolactam
Price:	$ 0.73/kg

Benzoic acid is manufactured mainly by liquid-phase air oxidation of toluene (110–120° and 2–3 bar) in the presence of cobalt salt catalysts.

The Amoco process is also used to oxidize toluene to benzoic acid. Amoco uses a combination of cobalt and manganese acetate in glacial acetic acid. Cocatalysts such as ammonium bromide or $CoBr_2$ are also used.

Decarboxylation of phthalic anhydride with steam is another process for the manufacture of benzoic acid. However, this process is not economically competitive with the direct oxidation of toluene.

Benzoic acid is used in the manufacture of phenol. Secondary products derived from benzoic acid include plasticizers (glycol dibenzoate) for vinyl resins; sodium benzoate, a food preservative; and benzoyl chloride, an intermediate in the manufacture of benzoyl peroxide. Benzoic acid is also used by Snia in Italy to produce caprolactam (see p. 142).

5. Dinitrotoluene

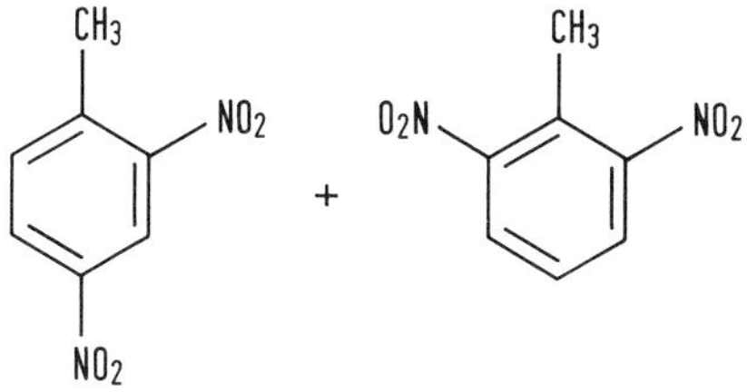

Consumption (1979):	United States	197,000 tons
	Western Europe	184,000 tons
	Japan	36,000 tons
Production Process:	Nitration of toluene	
Major Uses:	The only use is tolylene diisocyanate	
Price:	$ 2.03/kg (tolylene diisocyanate)	

The continuous nitration of toluene to give an 80:20 mixture of 2,4- and 2,6-dinitrotoluene is conducted on a large scale because the derived diisocyanates are the key building blocks for flexible polyurethane foams, which are widely used in furniture, bedding, and transportation applications. The reduction of the dinitrotoluenes to the corresponding diamines, which is usually conducted at 100°C and 50 bar using a Raney nickel catalyst, followed by continuous vapor-phase phosgenation produces tolylene diisocyanates (TDI). The total consumption of TDI in the United States in 1985 was 209,000 tons.

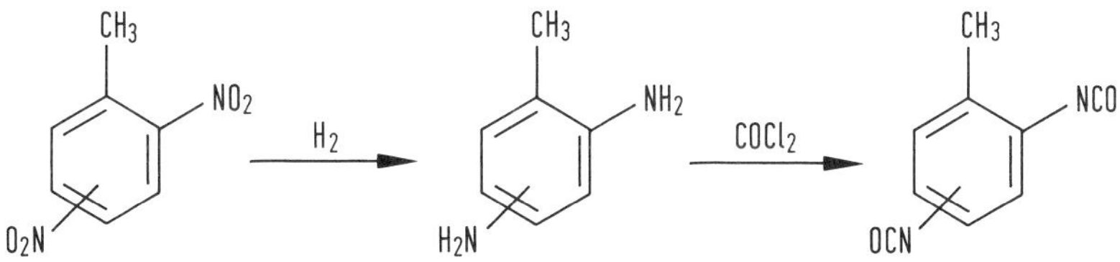

TDI

Recently a nonphosgene route to TDI has been explored by all producers. In this process the isomeric mixture of dinitrotoluenes is reacted with carbon monoxide in the presence of palladium or selenium catalysts to give TDI. A variation of this method involves reaction of the dinitrotoluenes with carbon monoxide in methanol to produce the dimethyl carbamates, which upon thermolysis yield TDI.

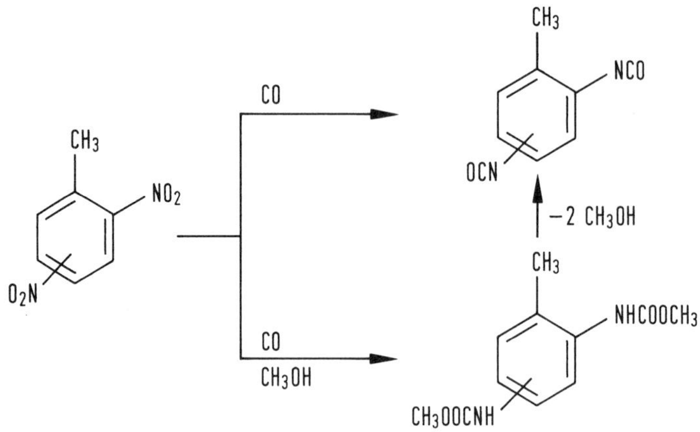

Although many companies have piloted this technology – notably Olin, Bayer, Mitsui Toatsu, and Mitsubishi – no commercial plant has yet been built.

Tolylene diisocyanate is used as the main building block for flexible polyurethane foam. The comonomers used are polyether polyols, mainly glycerine-initiated polypropylene oxides. The consumption of flexible foam in 1983 is summarized in Table 34.

Table 34. Consumption of Flexible Polyurethane Foam, 1983 (tons)

United States and Canada	554,000
Western Europe	730,000
Asia	240,000

Chapter IX. Xylene Derivatives

1. Introduction

In order to use the xylenes from the BTX fraction obtained in the catalytic reforming of naphtha as polymer intermediates, it is necessary to separate the isomers. Separation of the *p*-xylene from the C_8 fraction is conducted by distillation and fractional crystallization. For example, the higher boiling *o*-xylene is removed by fractional distillation, and the mixture of *p*-xylene and *m*-xylene is separated by several fractional crystallization steps. The higher melting *p*-xylene can be obtained in 99.5% purity, and the *meta*-isomer is mainly reconverted to *o*- and *p*-xylene because it has only a limited market. The isomerization of *m*-xylene can be conducted with Friedel-Crafts catalysts in the liquid phase or in the vapor phase in the presence of hydrogen. Amoco has developed a process to oxidize *m*-xylene-rich mixtures to isophthalic and terephthalic acids, which are separated at that stage. Both dicarboxylic acids are important synthetic fiber intermediates.

All three xylenes are important raw materials for industrial polymers. *o*-Xylene is mainly oxidized to phthalic anhydride, which is used widely in unsaturated polyester resins; *p*-xylene is oxidized to terephthalic acid, the basic building block for synthetic polyester fibers and aramid fibers; and *m*-xylene is oxidized to isophthalic acid, which is used in the manufacture of aramid fibers. Functionalization of the xylene isomers by air oxidation is an economically attractive process.

Naphthalene is also obtained from crude oil fractions by liquid-liquid extraction or adsorption. It is still being used for the manufacture of phthalic anhydride, although the air oxidation of *o*-xylene is an economically more attractive process.

2. *o*-Xylene

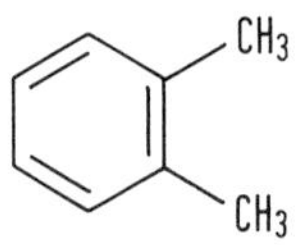

Consumption* (1984):	United States	312,000 tons
	Western Europe	743,000 tons
	Japan	190,000 tons
Production Process:	Separation from mixed xylenes made in oil refineries or in steam cracking olefin plants	
Major Uses:	Phthalic anhydride (95%)	
Price:	$ 0.40/kg	

The amount of *o*-xylene isolated depends entirely on the consumption of phthalic anhydride. Large quantities of *o*-xylene in the mixed isomer stream are used in gasoline or are isomerized to *p*-xylene. Phthalic anhydride is a versatile building block for polyesters and polyurethanes.

Monochlorination of *o*-xylene yields the starting material for benzocyclobutene, a monomer for high-performance thermoset polymers.

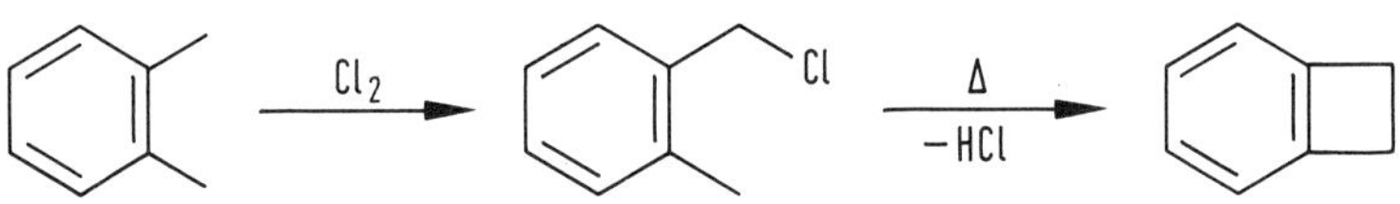

The thermal polymerization of benzocyclobutene involves reactive intermediates that undergo cyclopolymerization reactions (Diels-Alder type).

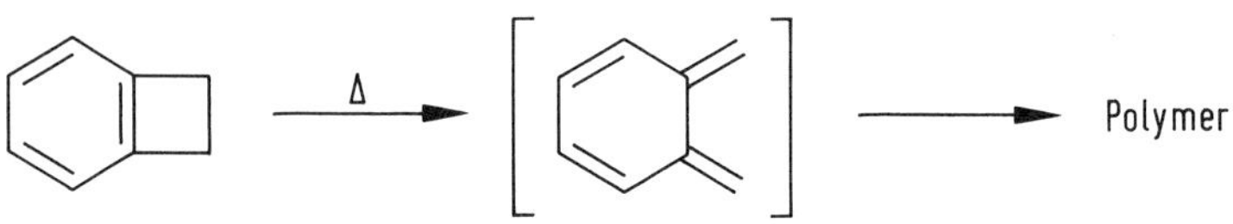

* Isolated *o*-xylene for chemical use.

3. Phthalic Anhydride

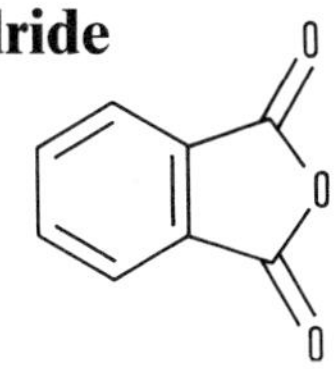

Consumption (1983):	United States	385,000 tons
	Western Europe	634,000 tons
	Japan	220,000 tons
Production Process:	Oxidation of *o*-xylene (71%); oxidation of naphthalene (29%)	
Major Uses:	Plasticizers (49%), unsaturated polyesters (24%), alkyd resins (20%)	
Price:	$ 0.85/kg	

About 71% of phthalic anhydride is produced by oxidizing *o*-xylene. This reaction is either conducted in the vapor phase using a fixed or a fluidized bed catalyst based on V_2O_5 or is a liquid-phase oxidation. The liquid-phase oxidation is conducted at 150°C using soluble acetates or naphthenates of cobalt, manganese, or molybdenum in combination with bromine as a promoter.

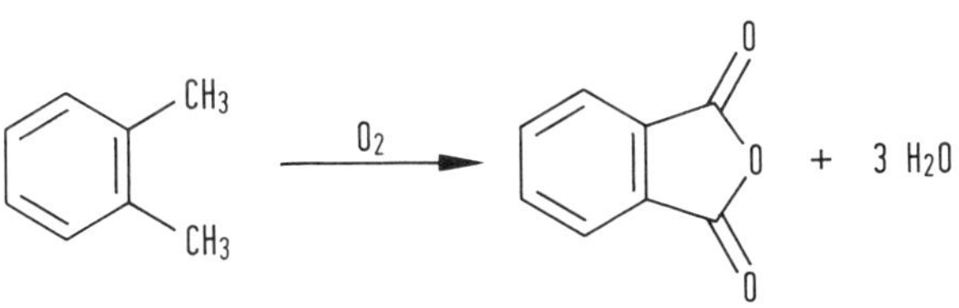

The air oxidation of naphthalene derived from petrochemical sources or coal tar is still being used to produce phthalic anhydride. The processes and catalysts are similar.

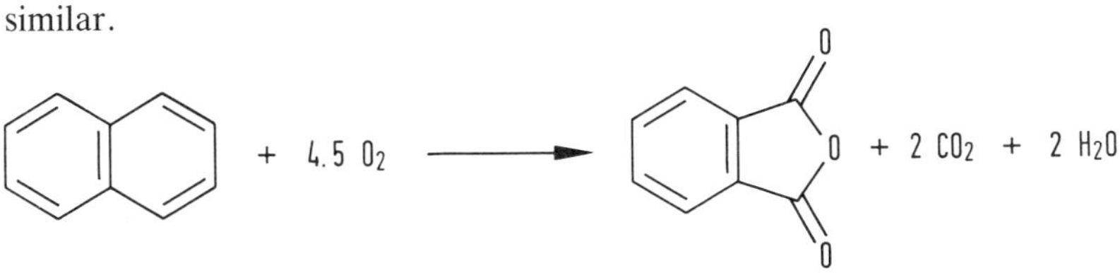

The selectivity to phthalic anhydride is on the order of 80–91%. The changing feedstock situation is shown in Table 35.

The most important industrial application of phthalic anhydride is the manufacture of plasticizers. The major diester used in this application is dioctyl phthalate. Dioctyl phthalate is produced from phthalic anhydride and 2-ethylhexanol. Another major application for phthalic anhydride is the manufacture of unsaturated polyesters, wich are used as binder resins for glass fibers. For example, SMC (sheet molding compound) is used to produce boat hulls and automotive body panels.

Table 35. Feedstocks for the Manufacture of Phthalic Anhydride (%)

Raw Materials	1966	1969	1975	1980
Naphthalene[a]	40	45	26	23
Naphthalene[b]	40	24	7	6
o-Xylene	20	31	67	71

[a] Based on coal tar,
[b] Based on petrochemicals.

Polyesters based on phthalic anhydride are also used in the manufacture of rigid polyurethane foams.

Nitration of phthalic anhydride followed by reaction with bisphenol A yields a dianhydride monomer used in the production of injection-moldable polyimides (Ultem).

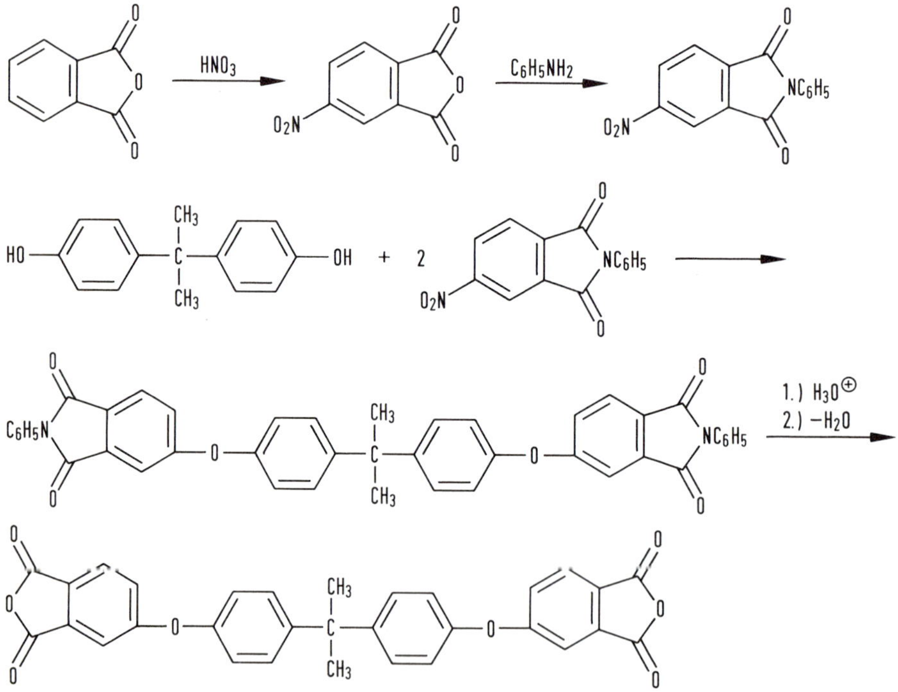

4. *m*-Xylene

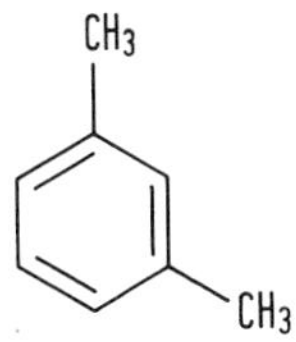

Consumption* (1980):	United States	41,000 tons
Production Process:	Separation from mixed xylenes	
Major Uses:	Isophthalic acid (88%), sulfophthalic acid	
Price:	\$ 0.68/kg	

The content of *m*-xylene in the mixture of isomeric xylenes is high, but most of the *m*-xylene is either returned to gasoline or isomerized to *p*-xylene. The isolated *m*-xylene is used mainly for the manufacture of isophthalic acid. Another use for *m*-xylene is in the manufacture of isophthalonitrile. Isophthalonitrile is used to produce *m*-xylenediamine, an epoxy curative. Phosgenation of *m*-xylenediamine affords *m*-xylene diisocyanate, a polyurethane intermediate. Other uses of *m*-xylene include *m*-toluic acid, which is used in the production of *N,N*-diethyl-*m*-toluamide, a mosquito repellant, and tetrachloroisophthalonitrile, an agricultural chemical. Sulfoisophthalic acid is used as a comonomer in the production of polyester fibers to improve their dyeability.

* Isolated *m*-xylene for chemical use.

5. Isophthalic Acid

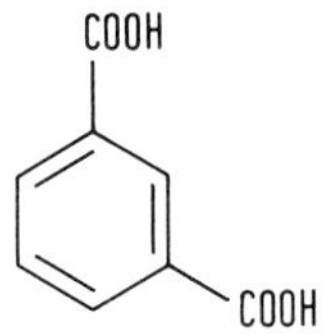

Consumption (1980):	United States	48,000 tons
	Western Europe	23,000 tons
	Japan	11,000 tons
Production Process:	Air oxidation of *m*-xylene	
Major Uses:	Polyester resins (54%), alkyd resins (23%), aramide fibers (9%)	
Price:	$ 0.95/kg	

Isophthalic acid is manufactured exclusively by air oxidation of *m*-xylene (Amoco process).

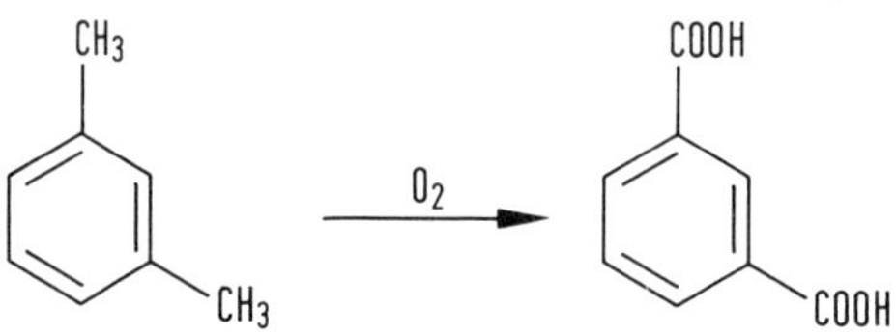

The oxidation is conducted in acetic acid using cobalt and manganese acetate as the catalysts. Isophthalic acid is a monomer for polyester resins and for high-performance aramide fibers. Dupont's Nomex aramide fiber is produced from isophthaloyl chloride and *m*-phenylenediamine.

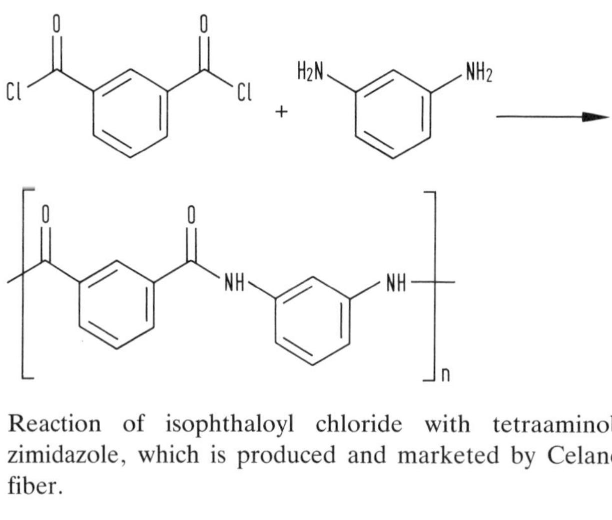

Reaction of isophthaloyl chloride with tetraaminobiphenyl yields polybenzimidazole, which is produced and marketed by Celanese as a high-performance fiber.

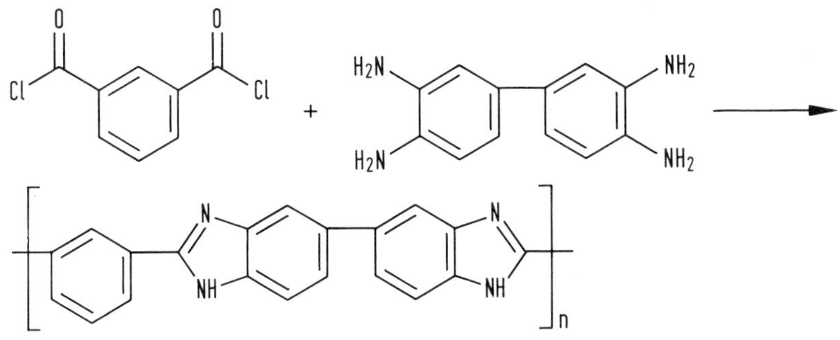

The ammoxidation of *m*-xylene produces isophthalonitrile, an intermediate for tetrachloroisophthalonitrile, a fungicide, and *m*-xylenediamine, the precursor for *m*-xylene diisocyanate.

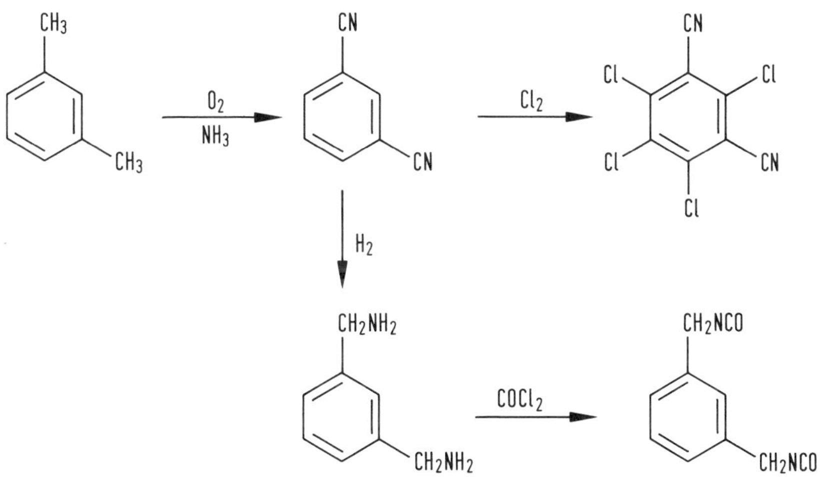

Further reduction of the aromatic ring in *m*-xylenediamine produces an aliphatic diamine, which after phosgenation to the corresponding diisocyanate provides a building block for color-stable polyurethanes.

6. *p*-Xylene

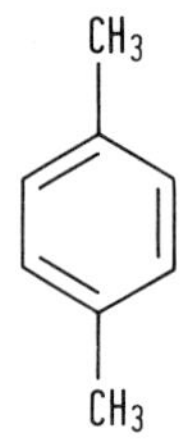

Consumption (1984):*	United States	938,000 tons
	Western Europe	884,000 tons
	Japan	722,000 tons
Production Process:	Separation from mixed xylenes obtained in petrochemical operations	
Major Uses:	Dimethyl terephthalate (55%), terephthalic acid (45%)	
Price:	\$ 0.40/kg	

Almost all of the produced *p*-xylene is used for the manufacture of dimethyl terephthalate and terephthalic acid, which are consumed almost entirely in the manufacture of polyester fibers, films, and molding compounds. Other commercial outlets for terephthalic acid include dimethyl tetrachloroterephthalic acid (a herbicide) and several high-performance condensation polymers.

* Isolated *p*-xylene for chemical use.

7. Terephthalic Acid and Dimethylterephthalate

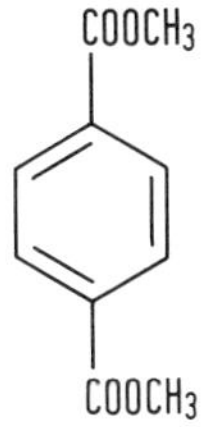

Consumption (1983):	United States	2,600,000 tons
	Western Europe	1,409,000 tons
	Japan	1,193,000 tons
Production Process:	Air oxidation of *p*-xylene	
Major Uses:	Poly(ethylene terephthalate) (90%) Poly(butylene terephthalate) (5%)	
Price:	\$ 0.85/kg,* \$ 0.77/kg**	

Terephthalic acid and its derivatives are used mainly as polyester intermediates. Polymers based on terephthalic acid that are resistant to higher temperatures have recently been developed. Initially, polyester fibers were produced from dimethyl terephthalate because this ester can be easily purified. However, Amoco has developed a process for the manufacture of pure terephthalic acid suitable for polymerization. Today about half of the total amount of polyester is produced from esterification of terephthalic acid with ethylene glycol instead of the original transesterification method.

Liquid oxidation of *p*-xylene involves *p*-toluic acid as an intermediate. Simultaneous esterification with methanol produces the methyl ester of toluic acid, which is readily oxidized to the monomethyl ester of terephthalic acid. Esterification with methanol yields dimethyl terephthalate (DMT).

* Terephthalic acid.
** Dimethylterephthalate.

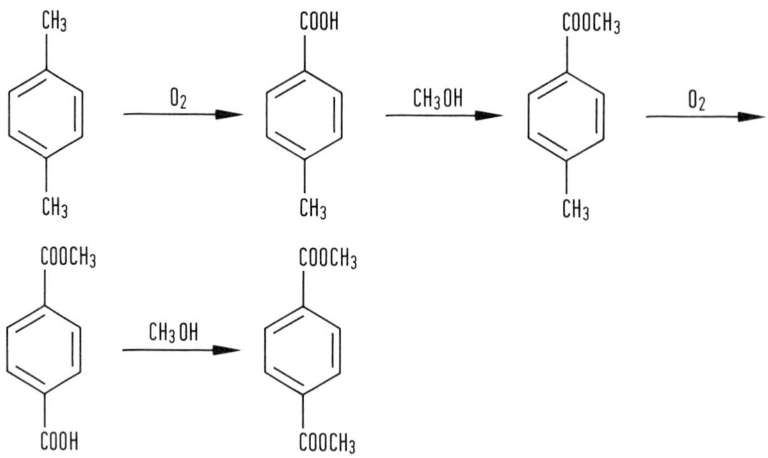

The oxidation of *p*-xylene to *p*-toluic acid is conducted with air at 140–170°C and 4–8 bar in the presence of cobalt and manganese salts of organic acids. Esterification with methanol is conducted at 250–280°C and 20–25 bar, and oxidation is continued simultaneously. The by-product, the monomethyl ester of toluic acid, is recycled. Several producers of DMT combine the air oxidation of *p*-xylene in the liquid phase and the esterification with methanol into a single step. The DMT thus obtained can be converted into fiber grade terephthalic acid by pressurized steam hydrolysis.

The inhibition of the oxidation of the methyl group in toluic acid can be overcome by adding bromine-containing oxidation promoters (NH_4Br, $CoBr_2$, $MnBr_2$). In the Amoco process the oxidation is conducted in 95% acetic acid, using a combination of cobalt and manganese acetate as the catalysts. The reaction occurs at 190–250°C and 15–30 bar, and it requires Hastelloy C equipment because of the corrosivenese of the catalyst system.

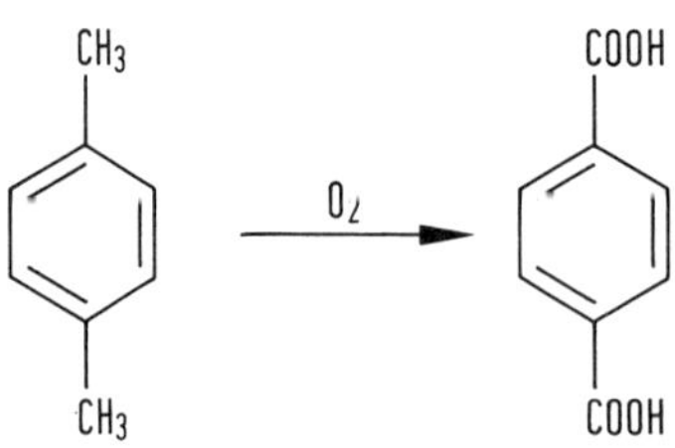

The *p*-xylene conversion is greater than 95%, and the selectivity to terephthalic acid is over 90%.

The Amoco process is also useful to oxidize other methyl group-containing aromatic compounds to the corresponding carboxylic acids. Typical examples include *m*-xylene to isophthalic acid, pseudocumene to trimellitic acid, and durene to pyromellitic acid. All of these functional carboxylic acids are used in the manufacture of thermally stable polymers as shown in Table 36.

Table 36. Monomers from Oxidation of Methylbenzenes

Substrate	Carboxylic Acid	Derived Polymers
p-Xylene	Terephthalic acid	Aramids (Kevlar)
	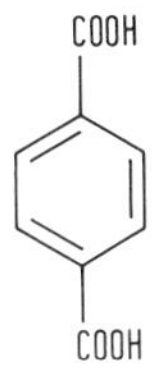	
m-Xylene	Isophthalic Acid	Aramids (Nomex), polybenz-imidazole
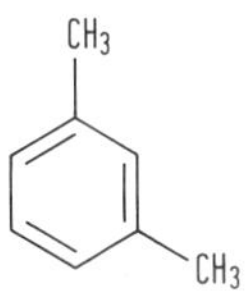	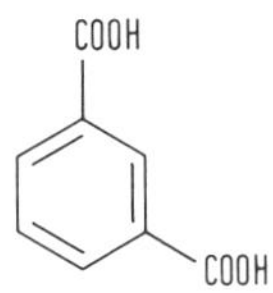	
Pseudocumene	Trimellitic acid	Poly(amide imide) (Torlon)
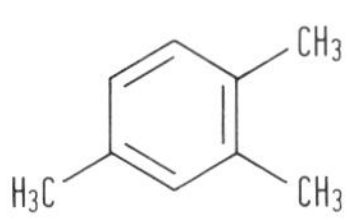	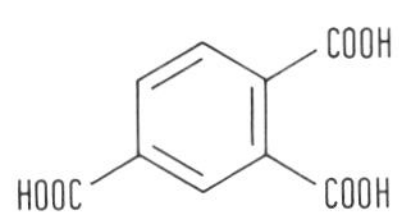	
Durene	Pyromellitic acid	Polyimide (Kapton, Vespel)
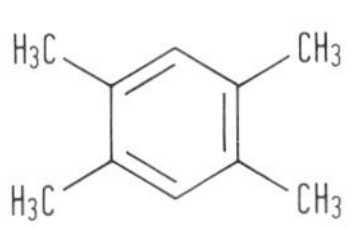	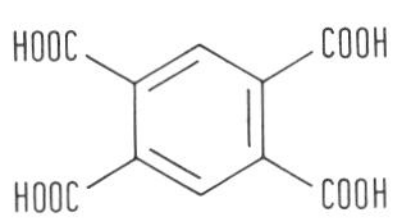	

The ammoxidation of *p*-xylene to terephthalonitrile has not been used commercially to produce terephthalic acid. However, Showa Danko in Japan produces terephthalonitrile and isophthalonitrile by ammoxidation of xylenes. The nitriles are reduced to the corresponding diamines, which are polymer intermediates and epoxy curatives. Phosgenation of the diamines yields the corresponding diisocyanates, which are polyurethane intermediates.

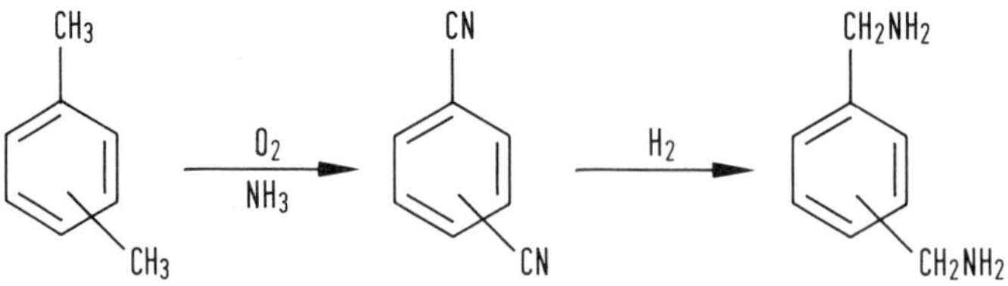

The Henkel processes involving the rearrangement of *o*-phthalic acid into terephthalic acid (Henkel I) or the reaction of the potassium salt of benzoic acid and CO_2 (Henkel II) have not been commercialized to produce terephthalic acid.

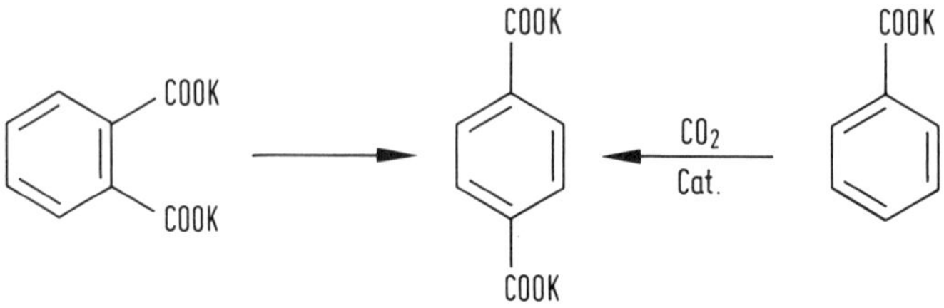

In the Henkel II process, potassium benzoate is disproportionated at 430–440°C in the presence of CO_2 and a cadmium or zinc benzoate catalyst to give dipotassium terephthalate and benzene.

The production of poly(ethylene terephthalate) is conducted by transesterification of dimethyl terephthalate or by esterification of terephthalic acid with ethylene glycol at 100–150°C and 10–70 bar in the presence of an acetate catalyst (Cu, Co, Zn, etc.). The intermediate bis(2-hydroxyethyl) terephthalate is further polymerized at 150–270°C with Sb_2O_3 as the catalyst. Either resulting polyester melts are spun into fibers or the polyester, after cooling, is granulated for used in injection molding and film applications.

The intermediate formed in the polycondensation of terephthalic acid with ethylene glycol, bis(2-hydroxyethyl) terephthalate, can be produced directly from terephthalic acid and ethylene oxide.

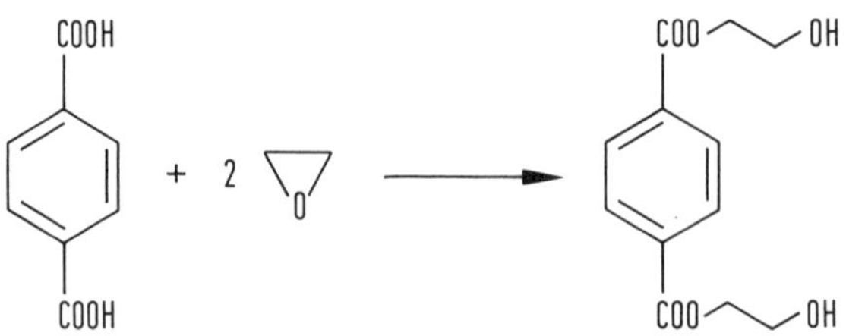

Dimethyl terephthalate is also used by Eastman Kodak to manufacture the comonomer 1,4-dimethylolcyclohexane used in the manufacture of Kodel fibers.

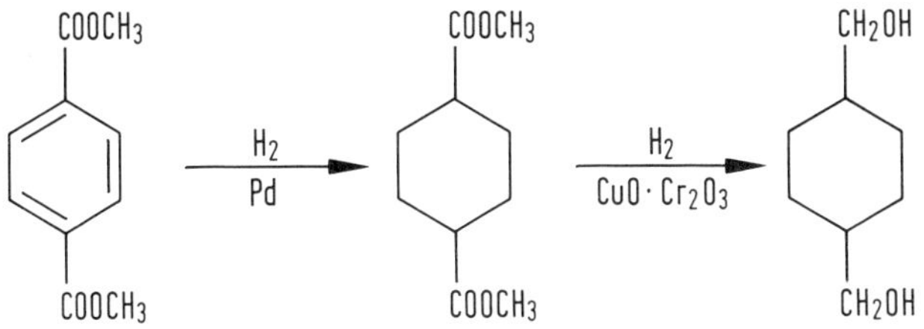

Data on the consumption of poly(ethylene terephthalate) (PET) and poly(butylene terephthalate), a molding resin obtained by reacting terephthalic acid with 1,4-butanediol, are shown in Table 31.

Table 42. Consumption of PET and PBT, 1983 (tons)

	PET	PBT
United States	2,028,000	32,000
Western Europe	875,000	14,000
Japan	638,000	31,000

About two-thirds of the PET is used in fibers; the remainder is used in bottles and films. Magnetic tape consumer products use the bulk of PET film, and molded bottles are the fastest growing market segment for polyester molding resins.

8. Naphthalene

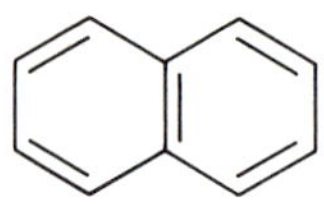

Consumption (1983):	United States	139,000 tons
	Western Europe	150,000 tons
	Japan	134,000 tons
Production Process:	Isolation from coal tar or petroleum refining streams	
Major Uses:	Phthalic anhydride (55%), Sevin (20%), α-naphthol (8%)	
Price:	\$ 0.50/kg	

About 60% of the naphthalene produced in the United States is obtained from coal tar; the remainder is isolated from petroleum refining streams. The major use of naphthalene, oxidation to phthalic anhydride, is declining because oxidation of *o*-xylene is more economical. α-Naphthol, derived from naphthalene, is the basic raw material for the insecticide Sevin. Union Carbide, the sole producer of Sevin, manufactures α-naphthol by reducing naphthalene to tetrahydronaphthalene, followed by oxidation to a mixture of 1-tetralol and 1-tetralone. Dehydrogenation of this mixture affords α-naphthol. Sevin is synthesized by reacting α-naphthol with methyl isocyanate. Since the accident involving methyl isocyanate in Bhopal, India, the future of Sevin has been in jeopardy. The current approach is to produce methyl isocyanate only as needed. Dupont has developed a new process for methyl isocyanate based on the oxidation of *N*-methylformamide (see p. 46). By generating methyl isocyanate *in situ*, the possibility of a major release is eliminated.

Chapter X. Monomers for High-Performance Polymers

1. Introduction

All of the high-volume polymers are based on simple and inexpensive raw materials. However, a small group of so-called high-performance polymers use higher-priced monomers, which are made from the basic raw materials by further chemical transformations. For example, the higher-priced engineering plastics as well as the more thermally stable polyimides are based on aromatic building blocks containing either ether, ketone, or sulfone groups because of the thermal and thermooxidative stability of these groups. Difunctional, trifunctional, and tetrafunctional monomers are used for the production of these high-performance thermoplastics. Also, self-reinforcing polymers based on the molecular alignment of liquid crystals are under active development. Liquid crystal polymers are based on biphenol, terephthalic acid, and *p*-hydroxybenzoic acid.

Photocrosslinkable polymers and photopolymerizable monomers are used in photofabrication (photoresist, photomilling) and in the manufacture of printing plates. Some of these polymers, such as poly(vinyl cinnamates), contain aromatic building blocks. Semiconductive polymers are also formulated from aromatic building blocks. Examples include poly(*p*-phenylene) and poly(phenylene sulfide) (PPS).

High-temperature adhesives are also made from aromatic building blocks. These polymers must join metals, ceramics, films, and combinations of these materials and must maintain the bonds under the severe conditions encountered in aircrafts, space vehicles, and missiles. Supersonic aircraft structures demand stability at 230°C for tens of thousands of hours. Missiles require mechanical performance for less than a minute, but temperatures can exceed 540°C. Epoxies are also used as matrix materials in aerospace applications. Lightweight prototype airplanes are already being built with airframes made entirely from fiber-reinforced plastics. Thermoset aromatic polymers based on benzocyclobutenes are also under development.

Heterocyclic polymers are another group of high-performance polymers. Typical examples include polybenzimidazoles and polybenzoxazoles. These materials are usually very difficult to process, but due to their liquid crystal structure in solution they can be converted to high-performance fibers. Wet spinning from highly polar acidic solvents such as concentrated sulfuric acid or polyphosphoric acid is the method of choice for the fabrication of these fibers.

2. Aryl Ethers

Aromatic ethers are manufactured from reactive halogen compounds and phenols. An example is 4,4'-diaminodiphenyl ether, the building block for high-performance polyimides (see p. 147). Reaction of 4,4'-diaminodiphenyl ether with pyromellitic acid dianhydride produces polyimide films or molding powders. More thermally stable photoresists are also formulated from 4,4'-diaminodiphenyl ether. These polymers have thermal stability up to 450°C, with a service temperature of 200°C. For example, reaction of pyromellitic acid dianhydride with β-hydroxyethyl methacrylate produces a pyromellitic acid diester, which upon reaction with thionyl chloride gives the corresponding dicarboxylic acid chloride. Further reaction with 4,4'-diaminodiphenyl ether affords the photoreactive polyamide. The remaining photoresist is heated at 300°C to achieve conversion to the polyimide with regeneration of the volatile alcohol.

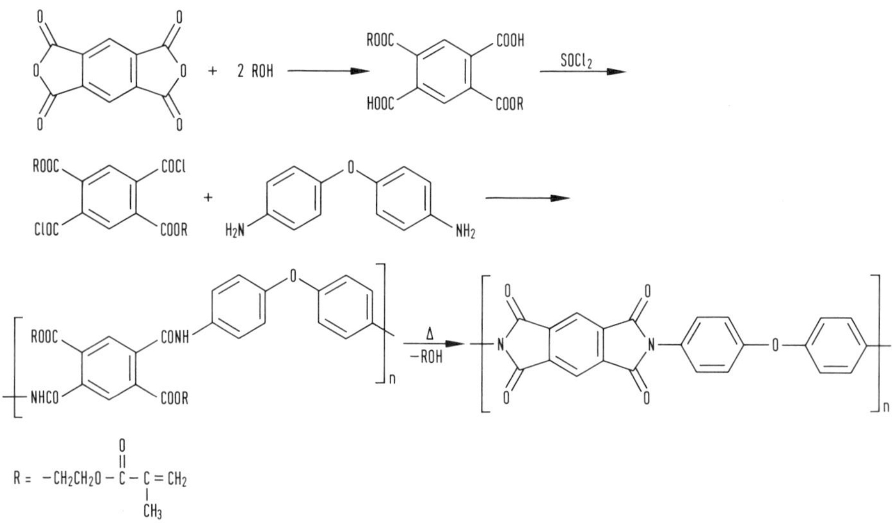

Another aromatic polyether is poly(phenylene oxide), which is obtained in the oxidative polycondensation of 2,6-dimethyl phenol (see p. 136). Aryl ether tetra-, tri-, and dicarboxylic acid monomers can be manufactured from *p*-chlorotoluene and *p*-cresol, followed by air oxidation of the methyl groups. The higher functional monomers are synthesized similarly from 3,4-xylenol and *p*-chlorotoluene or *p*-chloroxylene.

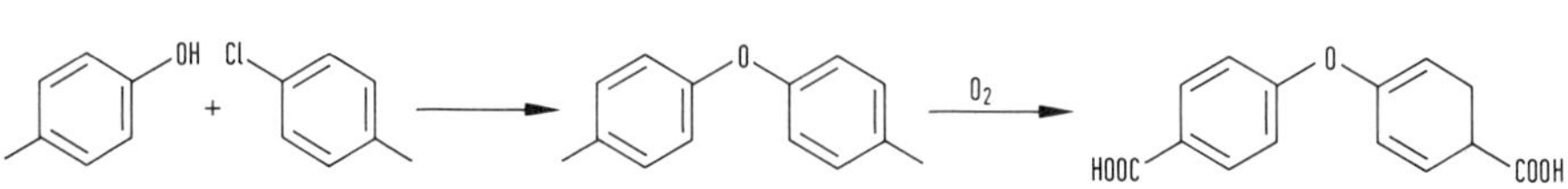

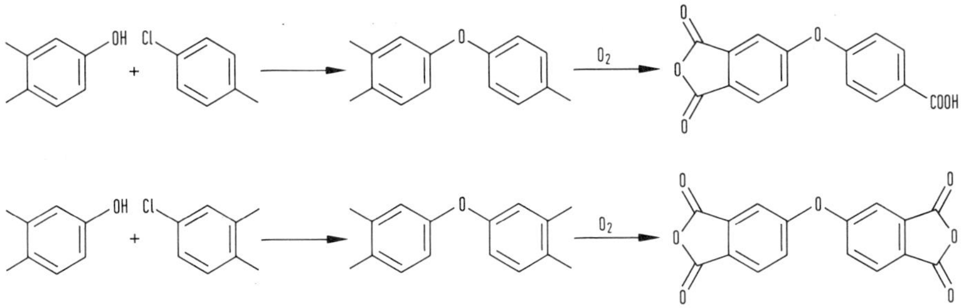

Upon reaction with diamines, the functional monomers, yield polyamides, polyamide imides, and polyimides, respectively. The molecular weight of the aryl ether monomers can be increased by increasing the number of aryl ether units. For example, reaction of *p*-cresol with *p*-dichlorobenzene followed by air oxidation affords the next higher homolog. Many variations on this theme have been explored in the laboratory, but commercialization has not yet been attempted.

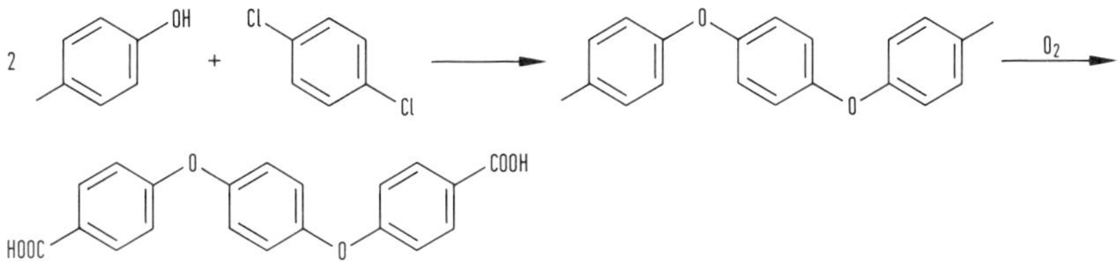

The polyamides obtained from these arylene oxides have a lower glass transition temperature (T_g), but hydrolytic stability is enhanced by lowering the number of polar amide linkages in the chain.

Bisphenol A is another useful building block for aryl ether polymers. Reaction of bisphenol A with 4-chloronitrobenzene followed by reduction of the nitro groups affords a diamine monomer for high-performance polymers.

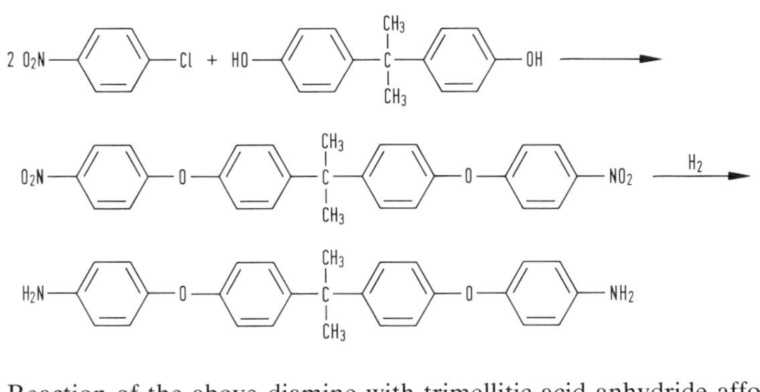

Reaction of the above diamine with trimellitic acid anhydride affords an injection-moldable poly(amide imide).

Reaction of 3-nitrophenylphthalimide with bisphenol A affords a dianhydride used in the manufacture of an injection-moldable polyimide (see also p. 170).

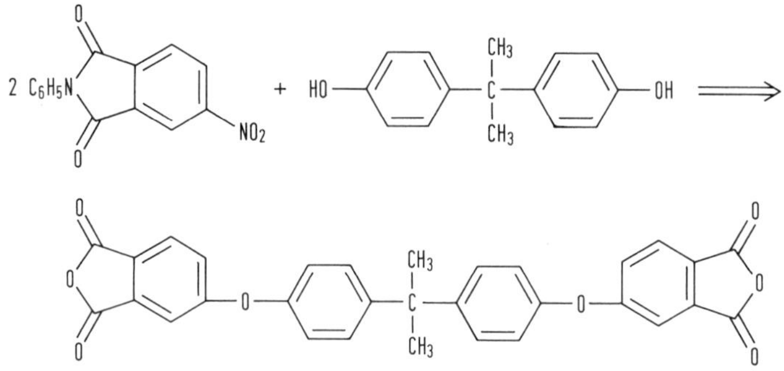

Aryl ether sulfones are also produced from bisphenol A and 4,4'-dichlorodiphenyl sulfone (see p. 154).

3. Aryl Ketones

Aryl ketones are also useful monomers for high-performance condensation polymers. One basic raw material for substituted benzophenones is 4,4'-diaminodiphenylmethane, readily available from aniline and formaldehyde (see p. 144). Oxidation of the methylene bridge, after protection of the amino groups, affords 4,4'-diaminobenzophenone. Diazotization prior to the oxidation allows introduction of other substituents. For example, 4,4'-difluorobenzophenone is the comonomer for PEEK (polyarylether ketone).

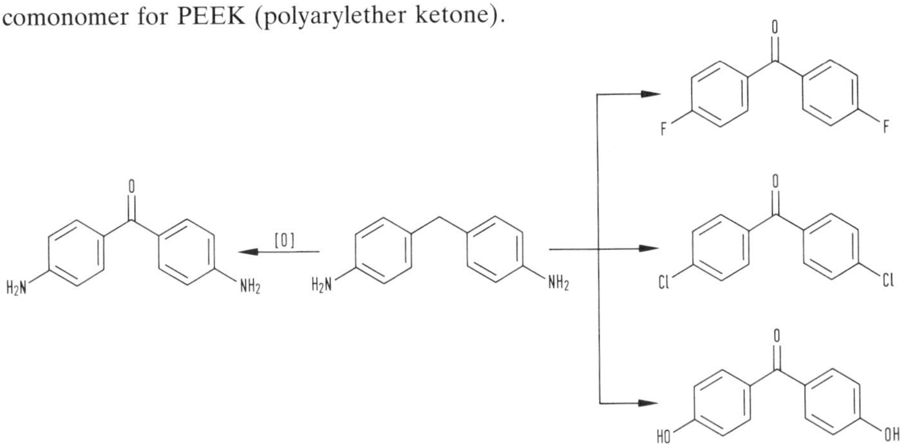

Another synthetic route to the halogenated benzophenones involves the Friedel-Crafts reaction of halobenzenes with phosgene. However, selectivity to the 4,4'-derivatives is low. Better selectivities can be obtained starting with 4-halobenzoyl chlorides.

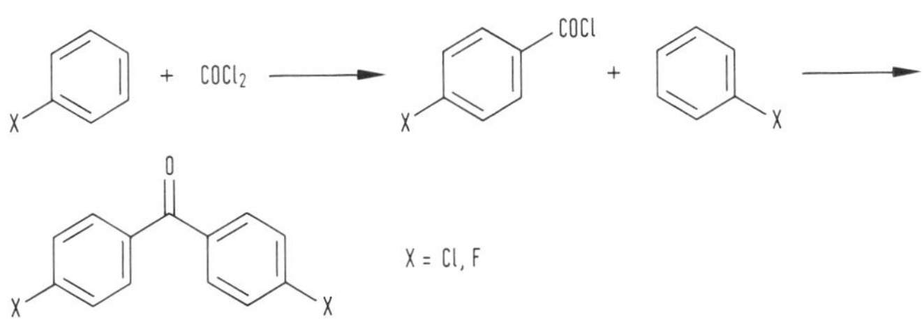

4,4'-Dihydroxybenzophenone can also be synthesized from 4-hydroxybenzoic acid and phenol, followed by a Fries rearrangement or by alkaline hydrolysis of 4,4'-dichlorobenzophenone.

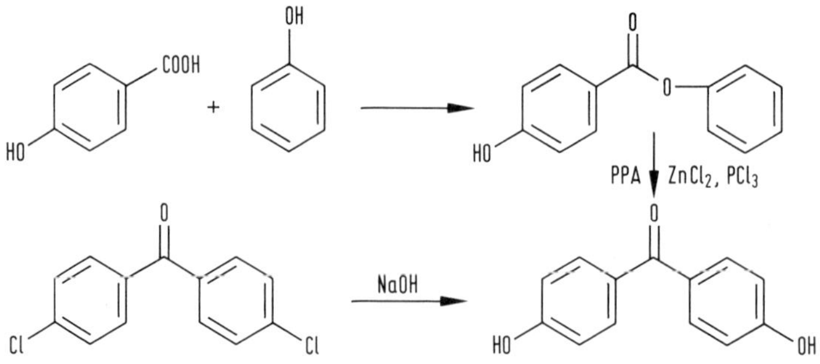

Poly(aryl ether ketones) are synthesized from 4,4'-difluorobenzophenone and the potassium salts of 4,4'-dihydroxybenzophenone and hydroquinone, respectively. Hydroquinone is less costly than 4,4'-dihydroxybenzophenone. Poly(aryl ether ketones) are currently marketed by ICI (PEEK).

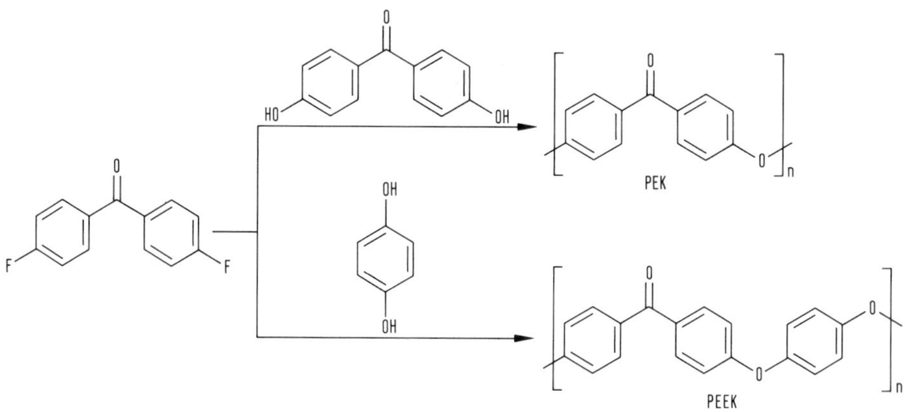

The polycondensation is conducted in diphenyl sulfone at 280–340°C to achieve sufficient solubility of the polymer. PEEK is a crystalline injection-moldable polymer with a T_m of 334°C and a service temperature of 220°C.

4,4'-Difluorobenzophenone is also the monomer of choice for poly(aryl thioether ketones). The comonomer in this case is inexpensive sodium sulfide.

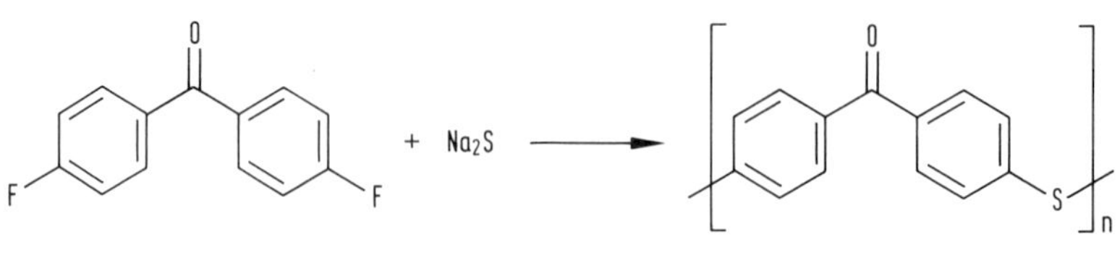

A tetrafunctional benzophenone derivative, benzophenone tetracarboxylic acid dianhydride (BTDA), is a monomer for polyimides. The basic raw material for BTDA is *o*-xylene. Reaction of *o*-xylene with acetaldehyde followed by nitric acid oxidation yields benzophenone tetracarboxylic acid, which is dehydrated to the dianhydride upon heating. A second approach to BTDA involves Friedel-Crafts reaction of *o*-xylene with phosgene to produce tetramethylbenzophenone (xylophenone). Subsequent oxidation of tetramethylbenzophenone with nitric acid and dehydration gives BTDA.

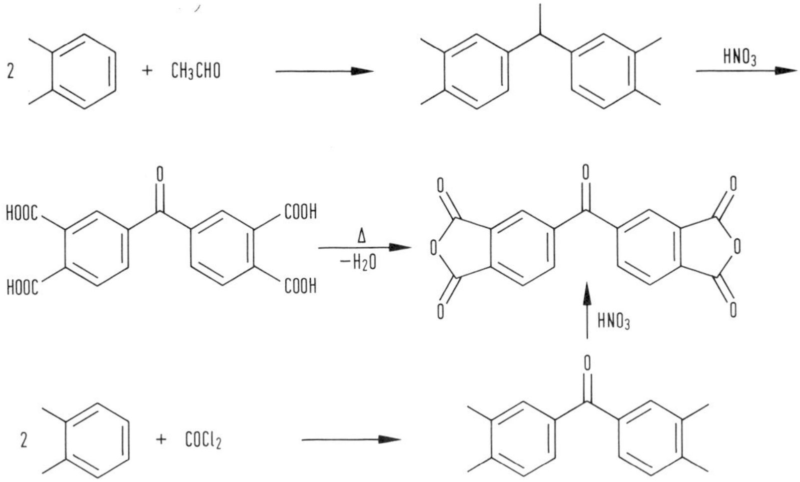

Polyimides are obtained from BTDA and *m*-phenylenediamine via the intermediate amide acid or directly from BTDA and diisocyanates.

Reaction of BTDA with 3,3'-diaminobenzophenone also produces polyimides with useful high-temperature properties. However, 3,3'-diaminobenzophenone is not readily available.

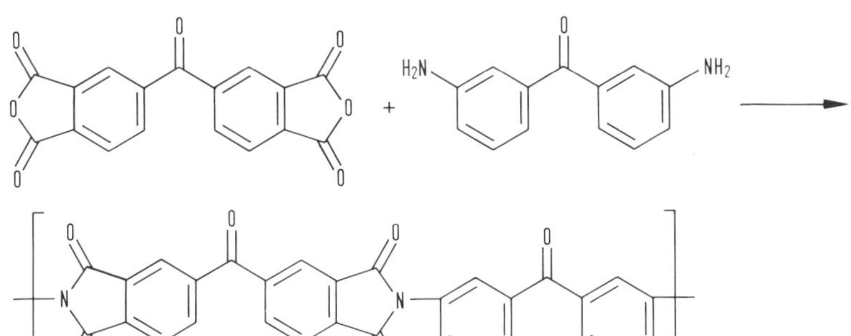

A combination of the two building blocks, aryl ketones and aryl ethers, produces monomers with relatively high molecular weights. The derived polymers have good processability but reduced thermal properties because of their lower glass transition temperatures. For example, reaction of 4,4'-dichlorobenzophenone with 3,4-xylenol and subsequent oxidation affords a dianhydride incorporating both structural features.

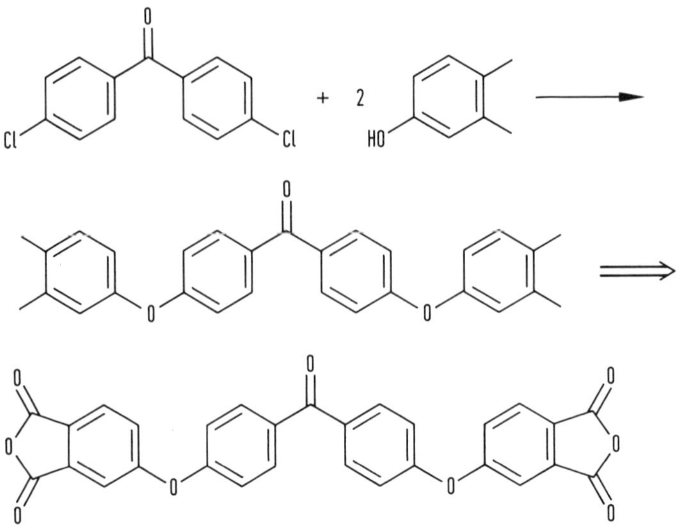

Reaction of this dianhydride with suitable diamines produces injection moldable polyimides.

4. Aryl Sulfones

The basic building block for polysulfones, 4,4'-dichlorodiphenyl sulfone, is produced by sulfonation of chlorobenzene using sulfur trioxide and excess diethyl sulfate, or sulfur trioxide and phosphorus pentoxide.

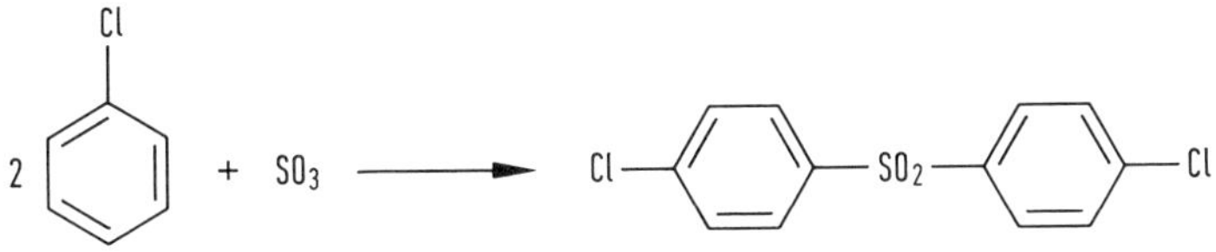

Alkaline hydrolysis of 4,4'-dichlorodiphenyl sulfone affords 4,4'-dihydroxydiphenyl sulfone in high yield.

Reaction of the sodium salt of bisphenol A with 4,4'-dichlorodiphenyl sulfone in dimethyl sulfoxide produces a polysulfone, which is currently marketed by Union Carbide under the trade name Udel.

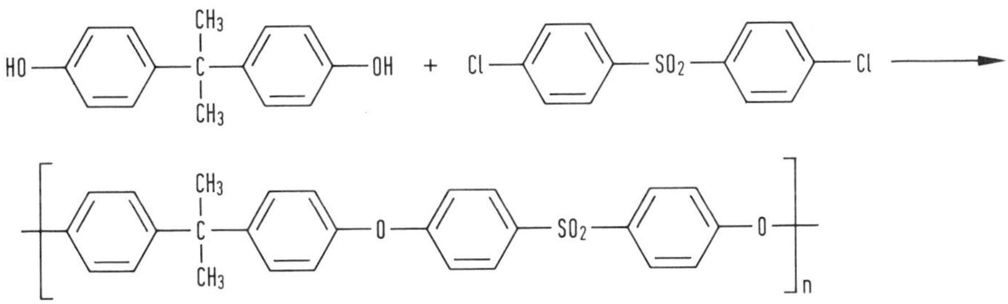

A similar aryl ether/sulfone type of polymer is obtainable from diphenyl ether. Sulfochlorination of diphenyl ether gives rise to the formation of a functional monomer, that can be homopolymerized under Friedel-Crafts reaction conditions.

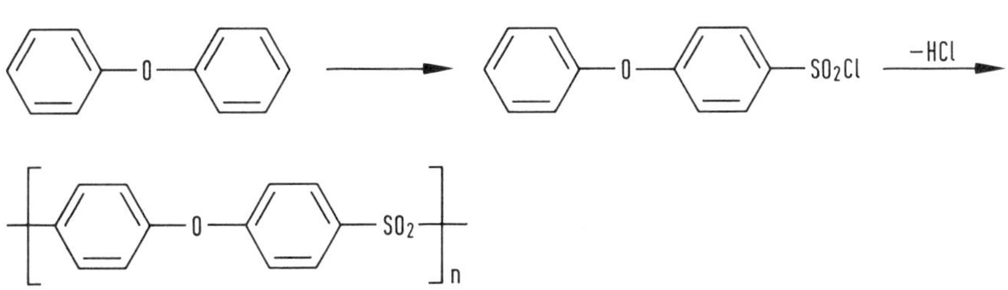

Another monomer can be obtained by sulfochlorination of biphenyl. These types of polysulfones are sold by 3M Corporation and ICI.

5. Functional Biphenyls

Two functional biphenyl derivatives are used as building blocks for high-performance polymers. 4,4'-Biphenol is a comonomer for liquid crystal–reinforced polymers, and tetraaminobiphenyl is used in the manufacture of polybenzimidazole fibers. Biphenyl itself is readily available from benzene, but selective functionalization is not easily accomplished. The benzidine rearrangement allows selective coupling of nitrobenzenes, but benzidine itself is no longer produced anywhere. Benzidine is a potent bladder carcinogen. In contrast, 3,3'-dichlorobenzidine is still commercially produced from *o*-chloronitrobenzene via the benzidine rearrangement. Reaction of 3,3'-dichlorobenzidine with ammonia yields tetraaminobiphenyl.

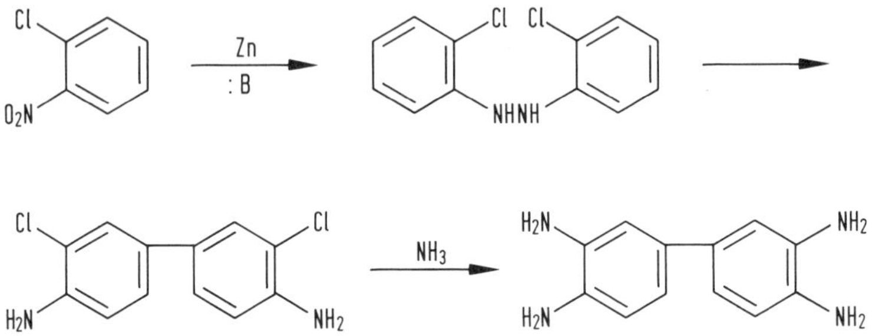

Polymerization of tetraaminobiphenyl with aromatic dicarboxylic acid chlorides gives polybenzimidazoles.

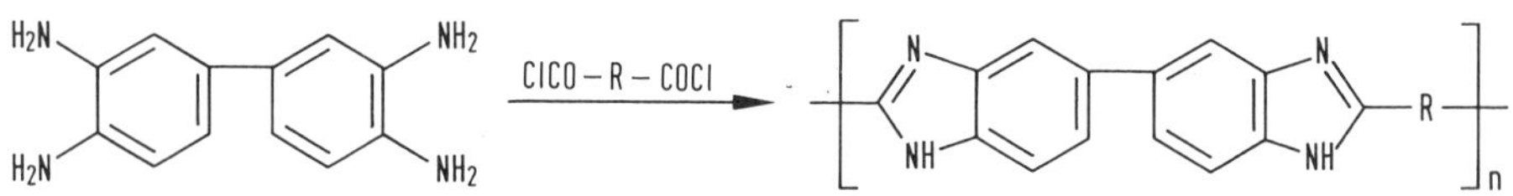

4,4'-Biphenol, the other important biphenyl building block, can be synthesized by tetrazotization of benzidine followed by hydrolysis. In view of the unavailability of benzidine, an indirect route is used for the manufacture of 4,4'-biphenol. Reaction of phenol with isobutylene affords 2,6-di-*tert*-butylphenol. Oxidative coupling of the phenol derivative produces 3,3',5,5'-tetra *tert* butyldiphenoquinone, which upon thermolysis in the presence of aluminum isopropoxide yields the desired 4,4'-biphenol. The by-product isobutylene can be recycled.

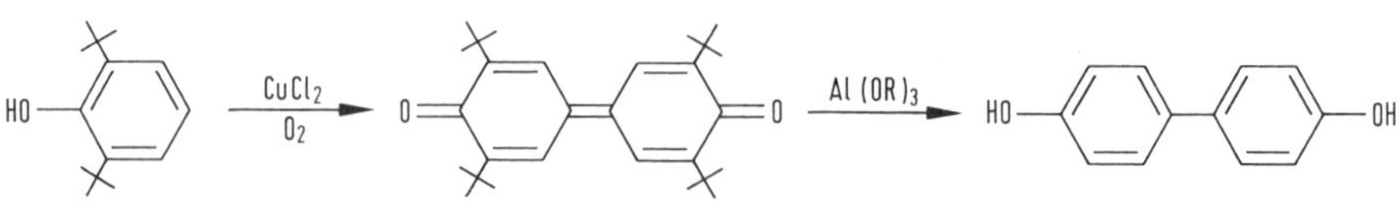

4,4-Biphenyldicarboxylic acid can also be obtained via tetrazotization of benzidine. However, no viable alternative route is available for this potentially useful monomer.

Chapter XI. Monomers and Polymers from Renewable Resources

The total amount of biomass formed by photosynthesis is estimated to be about 2×10^{11} tons/year, which is 20 times the total output of coal, oil, and natural gas. Crop-based processes already exist, and in some instances fatty acid manufacture and fermentation technology competes effectively with petrochemical-derived processes. In tropical countries with large areas of rapidly growing biomass, alcohol derived from a wide variety of fermentable polysaccharide-containing plants could be an important source for the synthesis of basic raw materials for polymers. For example, it is necessary to use an area the size of Belgium for the production of the sugarcane necessary to produce about 20% of the gasoline consumed in Brazil. Production of ethylene from ethanol leads to a further loss of 40% due to dehydration. Therefore, the replacement of the basic petrochemical feedstocks by renewable resources is not economically viable.

Cellulose is another basic product derived from chemical wood processing. Its world production equals more than 150 million tons per year. Agricultural residues and municipal wastes, which are predominantly paper, are also good sources for cellulose. During World War II, ethanol was produced from an acid hydrolyzate of wood cellulose. Microbial processes to convert cellulose obtained from municipal waste and papermill waste streams into fermentable sugars are already in the pilot plant stage. In Latin America, ethylene is already produced from ethanol by passing the vapors over a heated fixed-bed alumina catalyst. In the Unites States, initial efforts are under way to produce lignin and glucose/pentose syrup by steam explosion of wood followed by enzymatic degradation. Lignin is used in filler and adhesive applications, especially as a plywood adhesive and a particleboard binder resin. The glucose/pentose syrup is used for animal feed. Fermentation of the sugar residues may become economical by the end of this century. The development of plasmids to create alcohol-tolerant organisms would enhance the economic prospects of these processes. Commercial processes to produce benzene, phenol, cresols, and other aromatic compounds from lignin may also emerge. Dimethyl sulfoxide, a powerful polar solvent used in the manufacture of some condensation polymers, is already produced from the black liquor obtained in the Kraft pulping process. Heating the liquor converts lignin methyl groups into methyl sulfide, which is readily isolated and oxidized to dimethyl sulfoxide. Sulfonated lignins, a major component of the black liquor, can be recovered and used as dispersants, binder resins, sequesterants, and emulsion stabilizers. Crude calcium lignosulfonate is oxidized to produce vanillin.

All biomass consists largely of lignocellulose, which is produced and destroyed in nature at a rate of 100 billion tons/year. Algae may also fix up to 100 billion tons of carbon per year to produce a wide variety of polysaccharides. These algal polysaccharides contain a number of different sugar residues, especially glucose, galactose, mannose, and fructose and also contain uronic acid or sulfate ester residues.They are used mainly in the food industry. Since they are structurally related to chitin, they could conceivably be used as fillers in plastic composites. Chitin and calcium carbonate form a composite that makes up lobster shells.

Lignocellulose consists of three main components-cellulose, hemicellulose, and lignin. Cellulose, a polymer of β-D-1,4-linked anhydrous glucose units, comprises 40–60% of the cell walls of trees and plants. Hemicellulose, a branched polypentose with monomeric units consisting of xylose, arabinose, galactose, and rhamnose, accounts for 20–50% of the dry weight of plants. Cellulose and hemicellulose are imbedded in an amorphous layer of lignin, which accounts for the remaining 10–20% of the plant material.

The largest chemical use of wood is in the manufacture of wood pulp for paper production. Rayon, a cellulose fiber, is also produced from wood. In 1978 a total of 2,960,000 tons of rayon was produced worldwide. Natural fibers, such as wool, cotton, hemp, and sisal still play a major role in today's fiber consumption. In contrast, the conversion of lignocellulose into chemical feedstocks has not yet obtained major significance. The production of by-product acetic acid, methanol, and acetone in the manufacture of charcoal from wood has been discontinued. However, production of furfural from certain plant waste streams (oat hulls, maize ears, sugarcane stalks) is still ongoing (see p. 124). Lignocellulose is usually pretreated prior to further degradation. After chemical degradation, as in the production of wood pulp, waste products, such as hexoses and pentoses can be fermented to give ethanol, butanol, acetone, acetic acid, lactic acid, or 2,3-butanediol. Citric acid, gluconic acid, and itaconic acid are obtained by anaerobic fermentation of molasses. The fermentation process is carried out in large closed fermenters, and the mixture of acetone (30 parts), *n*-butanol (60 parts), and ethanol (10 parts) is separated by fractional distillation.

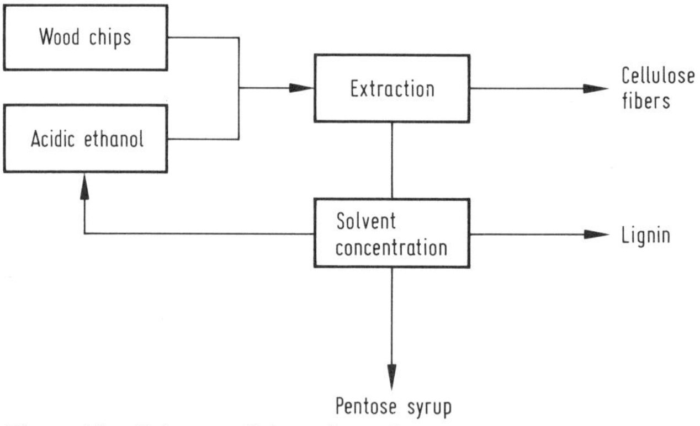

Figure 15. Solvent refining of wood.

In recent years, processes for biomass refining that focus on lignin isolation have been developed. In one process, wood chips are extracted with acidic or alkaline alcohol, which dissolves the lignin and hydrolyzes the hemicellulose to soluble sugars. The remaining cellulose is used as paper pulp. During alcohol stripping, the lignin precipitates and is collected for chemical use. The aqueous fraction is a solution of mixed sugars rich in pentoses that are used for cattle feed (see Figure 15). In a variation of this process, phenol is used as a solvent for lignin. Cracking of the lignin to produce benzene, phenol, cresols, and other aromatic compounds could generate the phenol solvent required for the process. However, the uses for the sugars from hemicellulose require very efficient removal of the phenol solvent.

A third process, which is in the pilot-plant stage, treats commercial wood chips with high-pressure steam. The release of the high pressure causes disintegration of the wood structure, which facilitates enzymatic hydrolysis. Part of the treated wood chips is used to produce cellulose enzymes, which are added to the remainder of the wood. The enzymatic hydrolysis of the cellulose produces 85% of the theoretically available cellulose, and the hemicellulose is readily hydrolyzed also. The lignin can be recovered from the residue by extraction with dilute alkali. Alternatively, the residue can be used directly as a plywood adhesive or as a particleboard binder resin (see Figure 16).

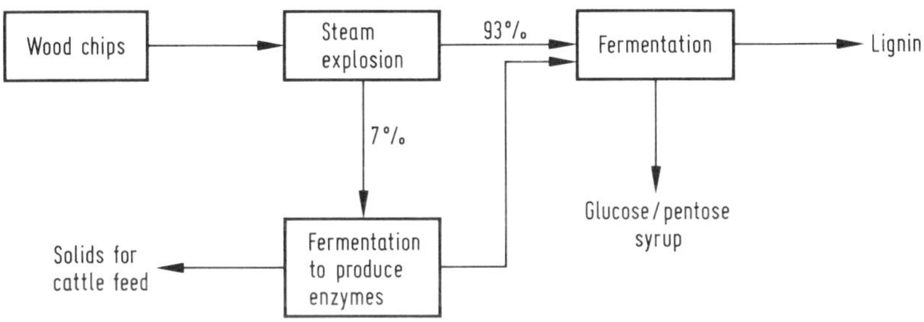

Figure 16. Steam explosion of wood.

An interesting polymer, poly(3-hydroxybutyrate), can be produced by fermentation of sugar or ethanol using *Alcaligenes eutrophus*. The cell wall of this microbe contains up to 80% poly(3-hydroxybutyrate), which can be obtained by extraction. The homopolymer can be used in the fabrication of biodegradable products. Under modified conditions, the same microbe produces a copolymer consisting of 3-hydroxybutyrate and 3-hydroxyvalerate units (see Figure 17).

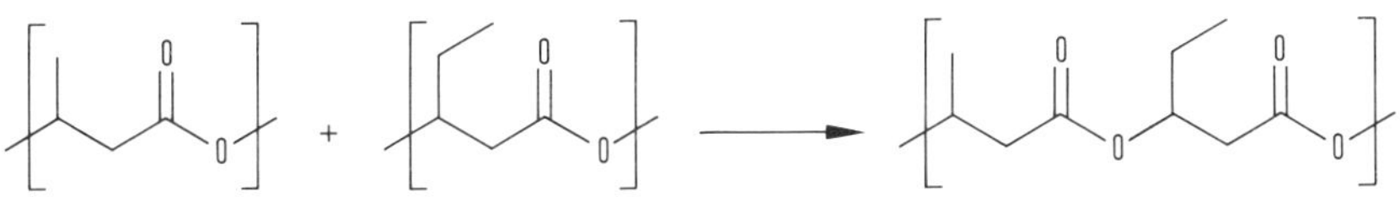

Figure 17. Copolymer derived from 3-hydroxybutyric acid and 3-hydroxyvaleric acid.

Biodegradation of the homopolymer as well as the copolymer is caused by common bacteria, fungi, and algae. In anerobic sewage, 100% degradation of 1 mm thick molded sheets occurs in 6 weeks. In anaerobic sewage and soil at 25°C, complete degradation takes 60–75 weeks. In seawater at 15°C, about 350 weeks is required.

The polymers are intended for packaging, for slow release encapsulation, and for the fabrication of implants that will be completely absorbed by the body.

Even high-performance polymers, such as polyphenylene, can be obtained by bacterial oxidation of benzene. The oxidation product, 5,6-dihydroxycyclohexa-1,3-diene, after esterification and pyrolysis in solution in the presence of radical initiators yields polyphenylene.

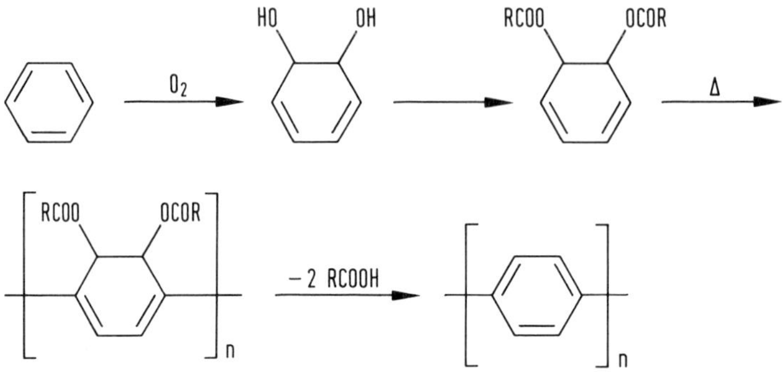

Polysaccharides are obtained not only from wood, but also from corn, wheat, other grains, and potatoes. These polymers are used extensively in the paper, foods, cosmetics, and textile industries. In 1983, in the United States alone, a total of 934,000 metric tons of polysaccharides was used (see Table 38).

Table 38. United States Polysaccharides Consumption, 1983 (tons)

Starches	805,454
Cellulose	61,363
Seed gums	21,818
Hydroxypropylguar	15,909
Seaweed extracts	11,363
Plant exudates	10,000
Xanthan	5,000
Pectin	3,180
Total	934,087

In the United States about 40% of the starch used in industry is modified either by partial depolymerization to dextrins or by chemical modification of the polymer chain. Modified starches include oxidized forms of starch, acetates, phosphates, amino alkyl ethers, and other ethers. They are used as paper and textile sizes and as binders for clay and other materials. Treatment of starch with ethylene oxide in an alkaline slurry produces hydroxyethyl ethers, which are useful in polyurethane applications. Also, graft polymerization of acrylonitrile to starch produces a highly absorbent material used in personal care products.

Sugar-derived polyurethane polyols are used in the manufacture of rigid polyurethane foam, wich is used as insulation in the construction industry. Reaction of sugars with propylene oxide yields polyols whose functionality depends on the starting sugar (see Table 39). These polyols are often tipped with ethylene oxide to obtain mainly primary hydroxy end groups. Glucose derived from starch is a pentol. High-pressure hydrogenation converts glucose into sorbitol, a hexol, and treatment of glucose with methanol/HCl affords α-methylglucoside with a functionality of 4. In contrast, sucrose obtained from sugarcane or sugar beets has a functionality of 8.

Table 39. Sugars Used in the Manufacture of Polyurethane Polyols

Sugar	Functionality
α-Methylglucoside	4
Glucose	5
Sorbitol	6
Sucrose	8

The initial high functionality in sorbitol can be reduced by dehydration as shown in Figure 18.

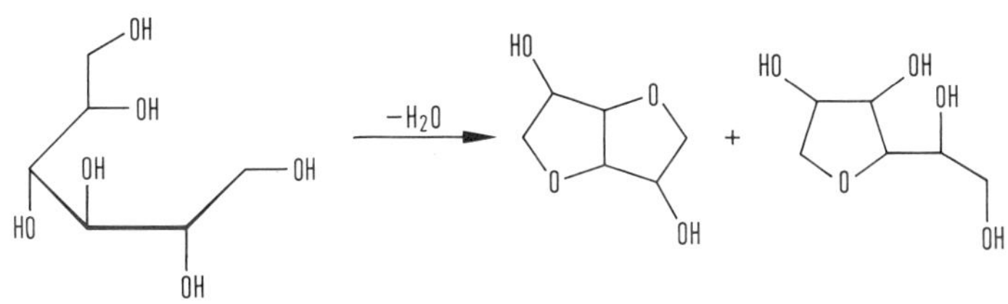

Figure 18. Dehydration of sorbitol.

Sorbitol is also used as a sweetening agent in foods and as a humectant stabilizer and emulsifier in drugs and cosmetics. It is used increasingly in toothpastes, especially gels, and in sugarless chewing gums. The United States consumption of sorbitol in 1984 was 120,000 tons.

Hydrolysis of starch affords glucose, which can be rearranged to fructose. The latter can be converted into 5-hydroxymethyl furfural, a useful comonomer for thermoset polymers. Condensation with phenol, urea, or melamine leads to interesting new condensation polymers (see Figure 19).

Treatment of starch with *Pullularia pullulans* produces polymaltotriose pullulan. Upon addition of water, this material behaves like a thermoplastic. It can be transformed into film and fibers, and it can be compression-molded. Its mechanical properties are similar to those of polystyrene.

Gums obtained from natural sources include water-soluble guar and locust bean gums from tree seeds and gum arabic from the sap of the acacia tree. The United States consumption of guar approaches 20,000 metric tons annually. Guar is used as a thickening agent for both processed foods and textile dyes. Locust bean gum also finds its major market in foods. Gum arabic is used in foods and lithography. In view of supply problems caused by adverse weather conditions, steps are being taken to change to seaweed extracts (sodium alginate, carrageenan). Synthetic polymers such as sodium carboxymethylcellulose are usually twice as expensive as natural guar gums.

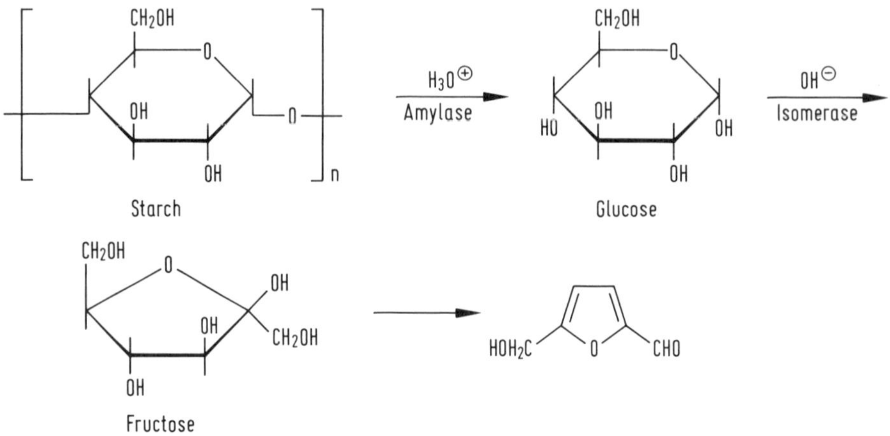

Figure 19. 5-Hydroxymethylfurfural from starch.

Oleochemicals are another class of raw materials obtained from renewable resources. The major raw materials for fatty acids include fats and oils and tall oil (rosin acids), a by-product of the production of wood pulp. Natural sources include vegetable oils; imported palm, palm-kernel, and coconut oils; tall oils; animal fats; and marine oils. The total market in the United States for oleochemicals in 1984 was about 900,000 metric tons with a value of about $ 1.2 billion. The use of oleochemicals is shown in Table 40.

Table 40. Use of Oleochemicals in the United States 1984 (total 900,000 tons)

Type of Product	Percent
Personal care	20
Polymer intermediates	18
Industrial lubricants	14
Coatings	10
Detergent intermediates	10

Crude fats and oils are usually refined by alkali treatment, which removes free carboxylic acids. The oxidative stability of oils and fats is improved by hydrogenation. Fatty acids are produced by high-pressure hydrolysis of fatty acid triglycerides, and glycerine is a valuable by-product. The glycerine obtained in this process competes effectively with the glycerine produced from petrochemical feedstocks (see p. 100).

Coconut oil is an important raw material for fatty acid production because of its high C_8-C_{14} content. For example, lauric acid (C_{12}) is obtained from coconut oil and palm kernel oil. Alternative sources for lauric acid are under development. While high-yielding palm kernel strains are beginning to show up in Malaysian palm plantations, cuphea, a family of oil seed plants, is considered an alternative source for lauric acid. Some cuphea species synthesize C_8 and C_{10} fatty acids that could potentially replace petroleum-based C_7 and C_9 fatty acids. Also, gene-splicing could eventually transform rapeseed into a lauric acid producer. Rapeseed produces longer chain fatty acids (C_{18} and up). Biotechnology is also used to increase the fatty acid content in plants. For example, sunflowers have been developed whose seeds contain up to 80% oleic acid. Older varieties produce seeds with about 40% of oleic acid. For longer chain fatty acids ($C_{20}-C_{22}$), marine oils are used.

The $C_{12}-C_{18}$ carboxylic acids are used for the manufacture of soaps. In recent years, synthetic detergents based on petrochemicals have largely replaced soap. However, some long-chain alcohols used as detergent builders are also produced from triglycerides.

Important nylon intermediates, such as azeleic acid (nylon-6,9) and ε-aminoundecanoic acid (nylon-11) are also produced from natural products.

Ozonolysis of oleic acid, which is obtained from liquid vegetable oils (cottonseed oil, soybean oil), produces a mixture of azeleic acid and pelargonic acid.

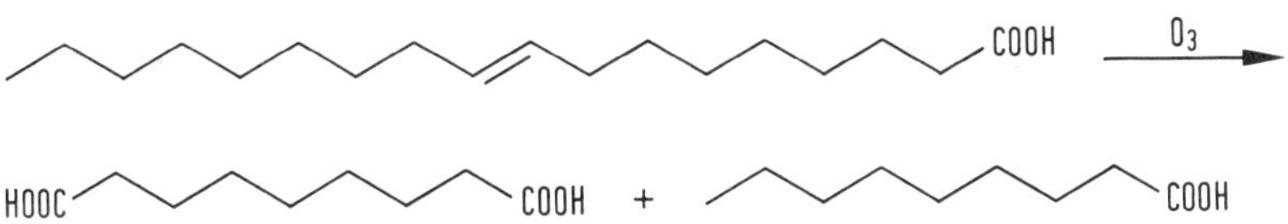

Azeleic acid can also be obtained from cyclooctadiene via ethoxycarbonylation. The resultant cyclooctene carboxylic acid affords azeleic acid upon treatment with sodium hydroxide at 320°C.

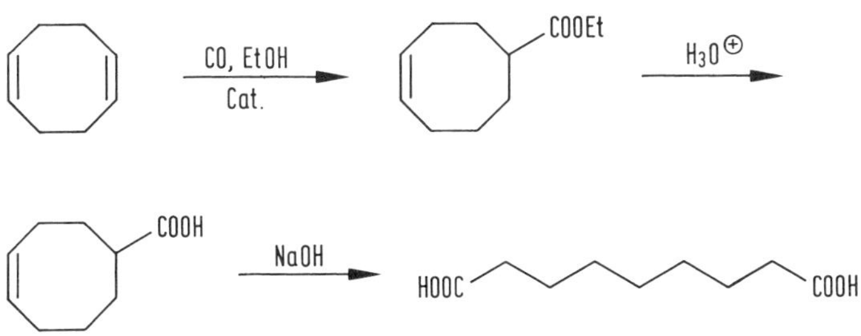

In the production of nylon-11, castor oil, available from Brazil and India, is transesterified with methanol to give glycerol and methyl ricinoleate. Pyrolysis of methyl ricinoleate at 450–500° produces methyl 10-undecylenate and heptaldehyde. Anti-Markownikoff addition of HBr to the double bond in undecenoic acid produces 11-bromoundecanoic acid which upon treatment with ammonia yields ε-aminoundecanoic acid, the monomer for nylon-11.

The dependence on natural products introduces uncertainties for future growth of nylon-11.

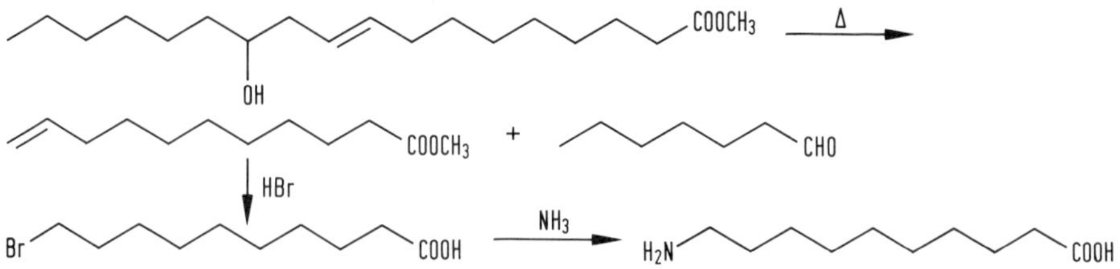

Ricinoleic acid is also the starting material for sebacic acid used in the synthesis of nylon-6,10. Sodium hydroxide fusion of ricinoleic acid at 250°C under pressure produces sebacic acid and 2-octanol.

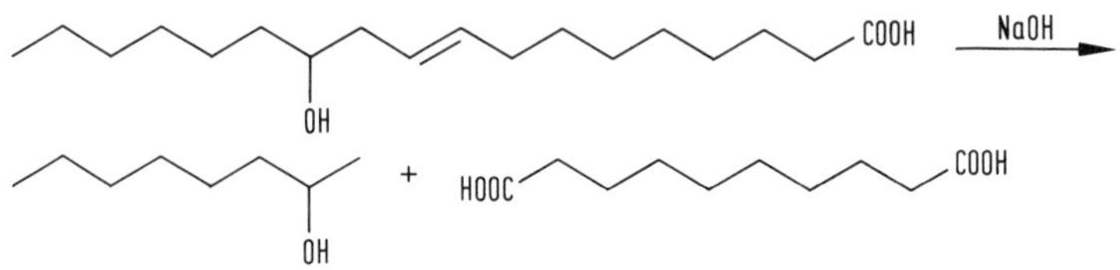

Other nylons, such as nylon-9 from soybean oil and nylon-13,13 from erucic acid, derived from rapeseed oil, have been manufactured on an experimental scale.

Natural rubber, obtained as an exudate from the species *Hevea brasiliensis*, is used extensively in the manufacture of rubber products. In 1979, 3 million tons of natural rubber was consumed in the world compared to about 5.9 million tons of synthetic rubber.

Poly-α-amino acids are excellent film and fiber formers, but they never became commercial products because of high prices. Their present methods of synthesis include fermentation of glucose, enzymatic conversion of various intermediates (fumaric acid, asparagine, etc.), extraction of proteins, direct synthesis, asymmetric synthesis, and chemical conversion of other α-amino acids. For example, α-amino acids upon treatment with phosgene form *N*-carboxyanhydrides, which undergo homopolymerization with optically active initiators to produce optically active polyamides) (see Figure 20).

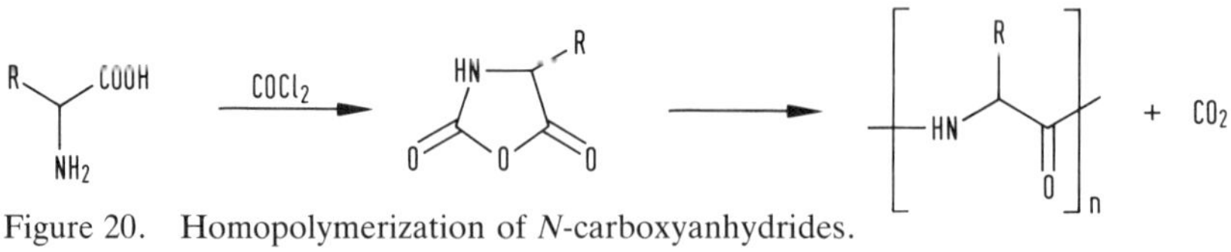

Figure 20. Homopolymerization of *N*-carboxyanhydrides.

Many biopolymers are produced by fermentation, and biotechnology will further improve today's fermentation technology. For example, hyaluronic acid, used as a buffer in eye operations, is currently obtained by extraction from cockscombs. Efforts are being made to develop bacterial routes to this biopolymer. Also, many higher plants produce important industrial and medical materials.

Advances in biotechnology, especially in the culturing of plant cells and tissues, will provide new ways to commercially produce such products. Plant-derived chemicals are expected to play an important role in the development of new products for regulating plant growth and for insect and weed control.

It is unlikely that large-volume polymers, such as polyethylene, will soon be produced from biomass via ethanol because the petroleum feedstock would have to double in price and the cost of production of fermentation ethanol drop substantially for it to be economically attractive, However, many new and unique functional polymers could become available from renewable resources.

Chapter XII. Future Trends

Today's industrial polymers are based on petrochemical feedstocks. The present situation will continue well into the next century. Worldwide oil and gas exploration will keep up with our current energy needs. The increased use of new sources of energy will make the petrochemical feedstocks more readily available as basic building blocks for synthetic polymers. Coal based C_1 chemistry is being actively explored in the laboratory, and new costeffective processes for basic raw materials based on synthesis gas are already in operation. Further refinement of the conventional process technology is in progress. The main emphasis is on the development of energy-efficient waste-free processes. Biomimetic processes with a high degree of selectivity are the target of today's process research. Modern metalorganic chemistry provides the sophisticated catalysts to achieve rapid reaction rates at lower temperatures. Some hydrogenation reactions that were historically conducted at elevated temperatures and pressures are now run at room temperature and atmospheric pressure. Sophisticated analytical methods and electronic process controls minimize process variability. The use of robotics in the chemical process industry reduces operator error. These trends will revolutionize the manufacture of the basic raw materials for industrial polymers.

Polymerization processes also benefit from the availability of higher purity monomers and vastly improved catalysts. Stoichiometry of the coreactants is of critical importance, especially in the manufacture of condensation polymers.

Processes leading to the manufacture of coproducts will be eliminated in time to avoid the need to balance the consumption of the derived products. An example is the production of phenol by air oxidation of cumene, which produces 0.6 lb of acetone per pound of phenol. Phenol and acetone are used in the manufacture of bisphenol A, a key monomer for polycarbonates and epoxy resins. However, numerous other growing uses of phenol will interfere with balancing the consumption of the coproduct acetone from the cumene process, A major use for acetone is in the formulation of solvent systems, and the emergence of solventless processes reduces the need for commercial solvents. In contrast, consumption of the coproduct *tert*-butanol, produced in the indirect oxidation of propylene, is not a problem because of its use as gasoline additive. Oxygenates, such as *tert*-butanol and MTBE, which can be readily made from *tert*-butanol and methanol, are used to improve the octane rating in gasoline. The ban on the use of tetraethyllead in gasoline in Europe has added to the demand of oxygenates.

The increased utility of macromolecules in today's world provides ample challenges to improve the processes for industrial monomers. Polymers are replacing metals in transportation to achieve fuel savings, and rust-free automotive body panels made from plastic products are already in use. Man-made fibers are used in more than 60% of the apparel and more than 75% of the home furnishings sold in the United States. Polymers are also used in increasing amounts in the packaging of foods and other consumer products. In addition to the traditional high-volume uses, new specialty applications are developing rapidly. Examples are polymeric membranes for water purification, heatstable polymers for use in oil exploration and in the electronics industry, high-performance composites for structural use in all forms of transportation, and special polymers for the health care business. Advanced composites, using carbon or aramide fibers in epoxy or polyimide matrix resins, are replacing metal in many applications. Performance advantages include weight reduction, corrosion resistance, and easy processing into intricate shapes. These types of composites are already used in aerospace applications. More mundane applications include fishing rods, golf club shafts, and tennis racquets. Fiber-reinforced plastic body parts are computer designed and implanted into the human body with outstanding results. Fiber reinforced poly(amide imide) (Torlon) has been used in the manufacture of an automotive engine. The world capacity for carbon fibers increased from about 25 metric tons in 1970 to about 5 500 metric tons in 1984.

More cost-effective fabrication technologies, employing thermoplastic matrix resins or rapid-curing thermosets are under development.

Modern living without polymers is unthinkable. Disposable diapers containing synthetic elastomers as moisture barriers and acrylic polymers to absorb moisture more effectively are marketed widely. Polymeric drug delivery systems and biodegradable implants are under development, and modern communication systems are using polymers as materials of construction as well as in the manufacture of integrated circuit chips. Electroconductive polymers will replace metal in electrical applications, and plastic minibatteries for motorcycles are under development. Liquid crystal copolymers provide self-reinforcing thermoplastics with unusual properties. Examples include copolyesters obtained from terephthalic acid, 4,4'-dihydroxybiphenyl, and *p*-hydroxybenzoic acid. The properties of liquid crystal-reinforced copolymers are superior to those of glass-reinforced materials.

To cope with the continuing demand for superior materials of construction, efforts are under way to improve the processes for the production of monomers as well as polymers. New and improved methods for recycling of polymers are also on the drawing board. Polyurethane scrap foam can be readily converted to polyols for use in the manufacture of new foam products. Even polyurethane automotive exterior body panels can be converted into useful polyols. Polyester bottle scrap is also converted into polyester polyols for use in the manufacture of polyurethane foam insulation boards. Other approaches are also being taken to upgrade bottle scrap into useful engineering plastics.

New polymer processing technologies are also providing challenges to develop entirely new families of liquid monomers to be used in the reaction injection molding of new materials of construction. Presently, RIM processes can only be used in addition polymerization processes. The challenge of the future is the manufacture of condensation polymers by polyaddition reactions. For example, reaction of liquid diketenes with macroglycols could produce polyester polymers using the RIM process. However, stable diketenes have not yet been synthesized.

Hot-melt stamping of reinforced plastics is another processing method that necessitates the development of new substrates. Hot-melt stamping requires less energy than injection molding of thermoplastic polymers or reaction injection molding of liquid monomers. It is also a substantially faster process.

The challenges in the manufacture of monomers for thermoplastic polymers relate basically to improvement of economics, reduction of energy, and elimination of by-products. Lower production costs can be achieved through reduction in the costs of raw material, labor, and overhead. Usually, the latter can be accomplished by utilization of the full plant capacity and by replacing plant operators with modern electronic monitoring devices. Reduction of energy requires the development of new process technologies that do not require solvents and reactions that are conducted at lower temperatures. The elimination of by-products requires processes that occur in almost quantitative yields.

Also, reaction of polymers with suitable organic chemicals to achieve functional properties will remain a priority in the years to come. An example are the photopolymers made from poly(vinyl alcohol) by reaction with cinnamoyl chloride in pyridine to give negative working photoresist resins. The attached cinnamoyl groups provide a site for photocrosslinking. In a similar manner, copolymers containing free hydroxyl or amino groups are reacted with 1,2-naphthoquinone-2-diazine-5-sulfonyl chloride to give positive working photoresists. Naturally occuring albumins, polylactides, and polysaccharides can be modified to give biodegradable polymers for medical applications. The reaction of linear copolymers containing hydroxyl or amino groups in the backbone structure is easily accomplished using many different reagents. The attached new moieties could provide a wide variety of functions, such as solubility in water, acids, or bases; electroconductivity; dyeability; and biodegradability.

In recent years, mixtures of polymers called alloys or polyblends are used to achieve maximum physical or chemical properties. For example, whereas thermoplastic polycarbonates undergo solvent stress cracking upon exposure to a variety of common solvents, this phenomenon can be greatly reduced by alloying polycarbonates with PBT or other polyesters. These alloys can be used to mold automotive bumpers that do not readily undergo stress cracking upon exposure to gasoline. Polymers can also be added as internal plasticizers to more brittle polymers to improve impact strength. The use of polymeric additives to improve fire resistance, oxidative stability, UV resistance, etc., is also preferred over the use of monomeric additives, which have a tendency to migrate from the polymer.

The use of new higher molecular weight monomers is another approach to achieve greater chemical stability in polymers. In these polymers, the number of linking polar groups is reduced, thereby improving the hydrolytic stability of the derived polymers. The synthesis of inexpensive higher molecular weight monomers will remain a challenge for the synthetic organic chemist. Ring-opening polymerization is another underutilized method for polymer synthesis. An example is the polymerization of caprolactam to nylon-6. This reaction can even be conducted in so-called hot RIM machines. Heating is required to convert the solid caprolactam into a liquid state suitable for filling the mold prior to polymerization. Nylon RIM automotive body panels are under development in the automotive industry. Of course, numerous other liquid monomers can be envisioned to be suitable for RIM processing.

Masked functionality provides a method to design novel one-component systems, that undergo rearrangement on heating to generate polymerizable groups. Addition of a catalyst to such a system can also lead to rearrangement of the masked monomer at room temperature.

The annual polymer consumption worldwide is in excess of 60 million metric tons, and increasing. A significant shift to production of the basic polyolefins at the wellhead could lead to improved economics for the commodity polymers. In the industrial countries, intense research efforts are being made to develop value-added specialty polymers. Examples are the segmented copolymers presently used to mold non-corrosive plastic cars. Superior hot-melt adhesives are also formulated to bond plastic parts used in the automotive and aerospace industries. The development of compact discs allows storage of enormous amounts of data; this development has already revolutionized the audio industry. The evolution from record to magnetic tape to compact disc has added new dimensions in fidelity and stability. The sales of digital audio discs made from polycarbonate are expected to grow from the 5.5 million discs sold in 1984 to 1.5 billion in 1994. Fiber optics, used increasingly in communication and biomedicine, need polymers as inner and outer jacketing.

The prospects for the future of industrial polymers are bright indeed, and rapid progress in the manufacturing technology of the required monomers is anticipated. The next century will be the age of polymers produced not only because of their mechanical properties, but also because of the other valuable functions they will perform in numerous applications, such as communications, transportation, housing, and biomedicine. Polymers will continue to be major contributors to the well-being of *Homo sapiens*.

Major World Producers of Monomers for Industrial Polymers

Monomer	Location	Company[a]	1985 Capacity[b] (1000 tons/yr)
Formaldehyde (37% soln.)	United States	Celanese	924
		Borden	709
		DuPont	668
		Georgia Pacific	538
		Monsanto	268
	Western Europe	BASF	1090
		CdF	424
		Montedison	380
		Degussa	324
		Bayer	247
	Japan	Mitsui Toatsu	299
		Mitsubishi Gas	279
		Daicel	150
Ethylene	United States	Shell	2254
		Dow	2022
		Union Carbide	1681
		Arco	1272
		Exxon	1181
		DuPont	1068
		Phillips	972
	Western Europe	Enichem Polimeri SpA	1290
		BP	1240
		Shell	1015
		Dow	940
		DSM	800
		Rheinische Olefinwerke (j. v. BASF/Shell)	650
		Veba	600
	Japan	Ukishima Petrochem.	750
		Mitsubishi Petrochem.	720
		Maruzen Petrochem.	490
		Sumitomo	430
		Sanyo Ethylene Co.	390
Ethylene oxide	United States	Shell	363
		Texaco	318
		Union Carbide	304
		ICI	227
		BASF	225
		Dow	211
	Western Europe	BASF	300
		BP	280
		Shell	260
		ICI	240
		Hoechst	160

[a] j. v. = joint venture.
[b] Changes have occured since 1985 due to mergers, acquisitions and leveraged buy-outs.

Major World Producers of Monomers for Industrial Polymers

Monomer	Location	Company[a]	1985 Capacity[b] (1000 tons/yr)
Ethylene oxide	Western Europe	Erdölchemie (j. v. Bayer/BP)	150
		Hüls	150
		Dow	130
	Japan	Nippon Shokubai	185
		Mitsubishi Petrochem.	146
		Mitsui Petrochem.	109
Ethylene glycol	United States	Union Carbide	886
		PD Glycol	281
		Shell	227
		Texaco	227
		Celanese	227
		ICI	215
		Dow	204
	Western Europe	BP[c]	265
		ICI	180
		BASF	160
		Hoechst	150
		Shell	130
		Dow	120
		Erdölchemie	110
	Japan	Nippon Shokubai	185
		Mitsubishi Petrochem.	146
		Mitsui Petrochem.	109
Vinyl acetate	United States	Celanese	420
		National Distillers	272
		Union Carbide	250
		DuPont	204
	Western Europe	Rhône-Poulenc	155
		Hoechst	150
		BP	90
		Wacker Chemie	80
	Japan	Kuraray	162
		Nippon Synth. Chem.	132
		Denka	120
Vinyl chloride	United States	B. F. Goodrich	908
		Dow	907
		Georgia Pacific	454
		PPG Industries	408
		Shell	381
	Western Europe	Enichem Polimeri SpA	800
		ICI	700
		Chloé Chimie SA	510
		Rovis-Rotterdamse (j. v. Akzo/Shell)	500
		Hüls	410

[a] j. v. = joint venture.
[b] Changes have occured since 1985 due to mergers, acquisitions and leveraged buy-outs.
[c] Includes the joint venture with Societé Francaise des Petroles.

Major World Producers of Monomers for Industrial Polymers

Monomer	Location	Company[a]	1985 Capacity[b] (1000 tons/yr)
Vinyl chloride	Japan	Kanegafuchi	290
		Kashima	270
		Ryo-Nichi Co. (Mitsubishi affil.)	220
		Sun Arrow Chem.	180
		Chiba VCM Co.	160
		Mitsui Toatsu	126
Propylene	United States	Shell	1522
		Exxon	1454
		Arco	850
		Standard Oil	836
		Gulf	793
		Texaco	602
		Union Carbide	547
		DuPont	545
		Dow	511
	Western Europe	Shell	805
		Dow	602
		BP (incl. j.v. with ICI)	590
		Erdölchemie	520
		Enichem Polimeri	520
		DSM	490
		Veba	372
	Japan	Ukishima Petrochem.	420
		Mitsubishi Petrochem.	340
		Tonen Petrochem.	327
		Sumitomo	264
		Osaka Petrochem.	240
		Maruzen Petrochem.	224
Propylene oxide	United States	Arco	644
		Dow	612
	Western Europe	Dow	375
		Arco	200
		Erdölchemie	135
		Shell	125
	Japan	Nihon Oxiran (50% Arco)	90
		Asahi Glass	85
		Mitsui Toatsu	36
Propylene glycol	United States	Dow	181
		Arco	113
	Western Europe	Dow	90
		Arco	60
		BASF	40
		Atochem	38
	Japan	Asahi	39
		AS Kasei (j.v. Asahi/Sumitomo)	12
		Showa Denko	10

[a] j.v. = joint venture.
[b] Changes have occured since 1985 due to mergers, acquisitions and leveraged buy-outs.

Major World Producers of Monomers for Industrial Polymers

Monomer	Location	Company[a]	1985 Capacity[b] (1000 tons/yr)
Acrylic esters	United States	Rohm and Haas	290
		Celanese	206
		BASF	104
		Union Carbide	100
	Western Europe	BASF	380
		CdF	150
	Japan	Nippon Shokubai	60
		Mitsubishi Petrochem.	34
		Asahi	20
		Ohita Chem. Co.	19
Acrylonitrile	United States	Monsanto	418
		Standard Oil	318
		DuPont	159
		Cyanamid	120
	Western Europe	Erdölchemie	280
		Monsanto	200
		Anic	160
		DSM	150
	Japan	Asahi	235
		Nitto Chem.	142
		Sumitomo	127
		Showa Denko	72
Epichlorohydrin	United States	Dow	191
		Shell	100
	Western Europe	Dow	80
		Shell	80
		Solvay	40
	Japan	Kashima Chem. Co.	24
		Osaka Soda Co.	22
		Showa Denko	8
Butadiene	United States	Shell	454
		Texaco	272
		Exxon	253
		Arco	204
	Western Europe	Erdölchemie	245
		Enichem Elastomeri	240
		Shell	190
		Dow	160
		Calatrava	132
		Bunawerke Hüls (j. v. Bayer/Hüls)	120
	Japan	Nippon Zeon Co.	140
		Japan Synth. Rubber	135
		Tobu Butadiene Co.	130
		Tonen Petrochem.	105

[a] j. v. = joint venture.
[b] Changes have occured since 1985 due to mergers, acquisitions and leveraged buy-outs.

Major World Producers of Monomers for Industrial Polymers

Monomer	Location	Company[a]	1985 Capacity[b] (1000 tons/yr)
Methyl methacrylate	United States	Rohm and Haas	300
		DuPont	109
		CYRO (j. v. Cyanamid/Roehm)	104
	Western Europe	Roehm	130
		ICI	105
		Degussa	45
	Japan	Mitsubishi Rayon	115
		Asahi	50
		Kyowa Gas	42
1,4-Butanediol	United States	DuPont	91
		GAF	52
		BASF	38
	Western Europe	BASF	70
		Hüls (j. v. with GAF)	30
	Japan	Mitsubishi Chem. Ind.	15
		Toyo Soda	6
Isoprene	United States	Goodyear	84
		Exxon	52
		Shell	45
		Arco	36
	Western Europe	Shell	75
		Compagnie Polyisoprene	40
	Japan	Geon Isoprene Co.	45
		Kashima	30
		Kuraray	30
Styrene	United States	Arco	1 145
		Dow	738
		Monsanto	681
		Hoechst	681
	Western Europe	Dow	658
		BASF	450
		Rheinische Olefinwerke	420
		Montedison	350
		Shell	325
	Japan	Mitsubishi Petrochem.	410
		Asahi	395
		Nihon Oxiran	225
		Nippon Steel	168
		Idemitsu	150
Phenol	United States	Aristech	263
		Georgia Pacific	234
		Shell	227
		Allied	227
		Dow	209
		General Electric	181

[a] j. v. = joint venture.
[b] Changes have occured since 1985 due to mergers, acquisitions and leveraged buy-outs.

Major World Producers of Monomers for Industrial Polymers

Monomer	Location	Company[a]	1985 Capacity[b] (1000 tons/yr)
Phenol	Western Europe	Phenolchemie	400
		DSM	255
		Montedison	215
		Shell	120
		Anic	90
	Japan	Mitsui Petrochem.	180
		Japan Phenol Co.	100
		Mitsui Toatsu	100
Adipic acid	United States	DuPont	500
		Monsanto	254
	Western Europe	ICI	300
		Rhône-Poulenc	230
		BASF	200
		Bayer	40
	Japan	Asahi	60
		Kanto Denka	10
Hexamethylene diamine	United States	DuPont	358
		Monsanto	182
	Western Europe	Rhône-Poulenc	180
		Polyamide Intermed.	90
		ICI	75
		BASF	55
	Japan	Asahi	30
Caprolactam	United States	Allied	227
		DSM	163
		BASF	163
	Western Europe	BASF	290
		DSM	220
		Bayer	160
		Montedison	102
	Japan	Ube	180
		Toray	135
		Mitsubishi Chem. Ind.	84
		Japan Lactam Ltd.	74
Bisphenol A	United States	Shell	158
		General Electric	100
		Dow	77
		USS	68
	Western Europe	Shell	90
		Bayer	60
		Dow	45
		Progelec	45
		General Electric	35
	Japan	Mitsui Toatsu	60
		Honshu Chem. Co.	30

[a] j. v. = joint venture.
[b] Changes have occured since 1985 due to mergers, acquisitions and leveraged buy-outs.

Major World Producers of Monomers for Industrial Polymers

Monomer	Location	Company[a]	1985 Capacity[b] (1000 tons/yr)
Isocyanates	United States	Mobay	226
		Dow	180
		BASF	101
		Olin	93
		Rubicon	74
	Western Europe	Bayer	288
		ICI	70
		Montedison	70
		Dow	50
		BASF	40
	Japan	Nippon Polyurethane	45
		MD Kasei (j. v. Mitsubishi/Dow)	36
		Sumitomo Bayer	36
		Mitsui Toatsu	18
		Takeda	20
Phthalic anhydride	United States	Monsanto	111
		Exxon	100
		Tenn-USS (j. v. Tenneco/USS)	95
		Stepan	77
	Western Europe	Hüls	200
		BASF	126
		Alusuisse	80
		UCB-Ftal	80
		BP	75
	Japan	Nippon Shokubai	99
		Kawasaki Kasei	72
		Japan Phthalic Co.	59
		Nippon Steel	52
Terephthalic acid	United States	Amoco	1396
		DuPont	862
		Hercofina	590
		Eastman Kodak	500
	Western Europe	ICI	456
		Dynamit Nobel	380
		Amoco	336
		Hoechst	295
		Montedison	140
	Japan	Mitsubishi Chem. Ind.	299
		Teijin	275
		Toray	260
		Mitsui Petrochem.	176

[a] j. v. = joint venture.
[b] Changes have occured since 1985 due to mergers, acquisitions and leveraged buy-outs.

Guide to Further Reading

Industrial Organic Chemistry, K. Weissermel and H.-J. Arpe, Verlag Chemie, Weinheim, New York, 1978.

Basic-Organic Chemistry, Part 5, *Industrial Products*, J. M. Tedder, A. Nechvatal, and A. H. Jubb, Wiley, New York, 1975.

Condensation Monomers, J. K. Stille and T. W. Campbell, Eds., Wiley-Interscience, New York, 1972.

Chemical Feedstocks from Coal, J. Falbe, Wiley, New York, 1982.

Future Sources of Organic Raw Materials, Chemrawn I, L. E. St-Pierre and G. R. Brown, Eds., Pergamon, New York, 1980.

Introduction to Industrial Polymers, H. Ulrich, Hanser, 1982.

Thermally Stable Polymers – Synthesis and Properties, P. E. Cassidy, Marcel Dekker, New York, 1980.

Polymer Synthesis (*Organic Chemistry*, Vol. 29, Parts I–III), S. R. Sandler and W. Karo, Academic, New York, 1974, 1977, 1980.

Chemistry and Application of Phenolic Resins, A. Knopf and W. Scheib, Springer-Verlag, New York, 1979.

Index